The Lacanian Delusion

FRANÇOIS ROUSTANG

Translated by Greg Sims

New York Oxford
OXFORD UNIVERSITY PRESS
1990

Oxford University Press

Oxford New York Toronto
Delhi Bombay Calcutta Madras Karachi
Petaling Jaya Singapore Hong Kong Tokyo
Nairobi Dar es Salaam Cape Town
Melbourne Auckland

and associated companies in
Berlin Ibadan

Published by Oxford University Press, Inc.
200 Madison Avenue, New York, New York 10016

Original French edition first published in 1986 as
Lacan, de l'équivoque à l'impasse, by Editions de Minuit.

Library of Congress Cataloging-in-Publication Data
Roustang, François.
[Lacan. English]
The Lacanian Delusion / by François Roustang;
translated by Greg Sims.
p. cm. —(Odeon)
Translation of: Lacan.
ISBN 0-19-506399-6
1. Lacan, Jacques, 1901–1981. 2. Psychoanalysis—France—History.
I. Title.
BF109.L28R6813 1990
150.19'5'092—dc20 89-26639 CIP

2 4 6 8 9 7 5 3 1

Printed in the United States of America
on acid-free paper

Acknowledgments

I owe a large measure of thanks to Christian Fournier, who gave his time so generously and served as an invaluable native guide through the more bristlingly impenetrable passages from Lacan's writings; to Rita Pazniokas, whose practiced editorial eye, sense of humor and English style led to rooting out a welter of not-quite-English barbarisms, to a translation no longer quite so manifestly stranded in the linguistic no-man's land of first and second drafts; to François Roustang, for his careful reading of the manuscript and for providing a much-needed computer in Paris; and finally to Josué Harari, for entrusting me with the project in the first place. Whatever Gallicisms, infelicities or inaccuracies remain are, of course, my own responsibility.

G.S.

Contents

Abbreviations

E: *Ecrits,* éditions du Seuil, Paris, 1966.

C: *Les complexes familiaux dans la formation de l'individu,* Navarin éditeur, Paris, 1984.

M: *Le Séminaire, livre II, Le moi dans la théorie de Freud et dans la technique de la psychanalyse (1954–1955),* éditions du Seuil, Paris, 1978.

P: *Le Séminaire, livre III, Les psychoses (1955–1956),* éditions du Seuil, Paris, 1981.

Q: *Le Séminaire, livre XI, Les quatre concepts fondamentaux de la psychanalyse (1964),* éditions du Seuil, Paris, 1973.

Eo: *Le Séminaire, livre XX, Encore (1972–1973),* éditions du Seuil, Paris, 1975.

Sc: *Scilicet,* n° 4, éditions du Seuil, Paris, 1973. (Journal of the Ecole freudienne of Paris, directed by Lacan).

Sci: *Scilicet,* n° 6-7, éditions du Seuil, Paris, 1976.

I: *Lacan in Italia, 1953–1978,* La Salamandra, Milan, 1978.

THE LACANIAN DELUSION

Psychoanalysis is distinguished by its extraordinary capacity for drifting and confusion, which makes its literature something that, I assure you, would not require a great deal of thought for it to be placed, in its entirety, under the heading of what we call literary madness.

Séminaire XI, p. 240 (1964)

When I said that they hate me, I meant that they don't suppose that I know [ils me désupposent le savoir]. And why not, if this turns out to be the necessary condition for what I have called reading? After all, what can I take for granted of what Aristotle knew? Perhaps my reading of him would be the better the less I took his knowledge for granted. Such is the condition for reading being rigorously put to the test, which I have no intention of dodging . . . People with good intentions—they're much worse than the ones with bad intentions.

Séminaire XX, p. 64 (1973)

1

Why Did We Follow Him For So Long?

This is a question repeatedly asked by some interlocutors, amazed to find criticisms being voiced by the very people whose indebtedness to him is incontestable, and a question I cannot avoid myself, since one must elucidate the reasons for such a long stretch of time being necessary for the flaws in the edifice to become clearly visible. Even if it is only ever possible to speak for oneself, it nonetheless remains the case that the personal experience in question was situated in a cultural context that I should try to evoke.

To begin with the obvious, Lacan was a figure wholly out of the ordinary; from the way he dressed, to the forms taken by his discourse, he could be compared to no other. In the first place, the color and cut of his shirts and suits, which must have come from the workshop of some Italian tailor, made him resemble a character from the *Commedia dell'arte* or a clown figure, as he himself acknowledged after screening the film version of his *Télévision*. Over the years, his seminar had become a spectacle, a show, and, if you did not always understand terribly well what he was proposing, you nevertheless remained spellbound by his capacity for (or his audacity in) shifting from one subject to another, for clarifying a topical point through some reference to a classical author, and for moving within the most diverse aspects of culture, as if he personally held all their secrets.

And how seductive he could be when marveling at some supposed discovery that you had made, to which he would call attention in digressing from a conversation, or after reading a text. How could any of his students resist when the man who apparently knew everything claimed to have been amazed by what you had said or written? It was as if your feeble intellect had somehow opened onto the possibility of momentary (and why not longer?) equality with a genius being fêted by seemingly everyone. Being singled out in that way would make even the soberest man's head spin. But, just as the Tarpeian Rock is itself not far from the Capitoline Hill, what a long fall it was when he lost interest, and you abruptly found yourself back in the midst of the collective stupidity! What to cling to then, once the sublime hopes he had inspired, and on which your very existence rested, had faded away, to be replaced by indifference, and, at times, contempt? Each disciple found his own solution: Flight, sometimes death, or, much more frequently, an even more passionate and blind attachment.

When founding the École Freudienne in 1964, Lacan had made use of his flair for large-scale politics, a flair he cunningly combined with the resources provided by psychoanalysis, in order to gather around him everyone who could be of some use to him. He was always able to come up with a strategy tailor-made for each individual. For example, a psychoanalyst who at the time held an important position in a remote province was hesitating in choosing sides: He was the object of numerous advances on Lacan's part, who confided in him, sought his advice, flattered him on occasion, and let him catch a glimpse of the role he would be able to play in the new organization; in short, Lacan treated him as an equal. However, once the analyst had finally made the right choice, he found himself on a decidedly different footing: Lacan imperiously summoned him, insisting that he come to Paris every week in order to undertake a joint review. You don't catch flies with vinegar . . .

One of Lacan's particular gifts lay in knowing how to adapt this strategy (a strategy that was well known all the same) to the style of his written and spoken interventions—a phenomenon that we shall have ample occasion to comment on later. He knew how to meet his listener or reader on the latter's own ground, with all his beliefs and prejudices; he would humor him for a while, indicating his agreement on many points. Then, once the other had been somewhat lulled into feeling secure, Lacan would exploit the authority he had acquired in order to impose his own views.

But it would be quite wrong to think of this authority as being somehow narrow, or petty. In the beginning, Lacan was surrounded by people who did not inevitably think as he did, and who did not imitate him in their work. Lacan was not unaware of this, nor did he refuse to accept it. In the early days of the École Freudienne, there were many who had been trained without having known him, and who were developing viewpoints markedly different from his. In those days, even his closest students had nothing to fear if they chose to follow their own itinerary and give their own personal touch to psychoanalysis. For years, this presented no problem. In one sense, no doubt, it is quite remarkable that we followed him for so long, since he was certainly not the only figure one encountered: There truly *were,* in the School, men and women who spoke frankly and who had their own style.

Later he even tempered the zeal of certain new lieutenants who were becoming intolerant of any deviation from obedience to the master. One day he reproached me for not having submitted an article to his journal *Scilicet.* When I responded by saying I was unable to write "Lacanian," he flew into a rage and invited me to write in my own way. Some time later, I sent him a text that he published, in spite of the reticence and objections of certain people responsible for the journal. Quite simply, I think that Lacan felt strong enough to sustain a diversity that not only did not overshadow him, but actually gave greater scope to his power.

In order to gauge the impact of this figure on French psychoanalysis, it should be remembered that, in its early days, psychoanalysis was the concern of only a very restricted group. Even if he wasn't the only person—far from it—behind its expansion, it was largely through Lacan that psychoanalysis made its way into the culture, and was able to attract into its sphere numerous people from other disciplines. As we know, Lacan established links between psychoanalysis and philosophy, ethnology, linguistics, mathematics, and even theology—hence the impression of his having produced, in an age of specializations, what had only been possible in previous centuries, namely a synthesis of all forms of knowledge, the reemergence of the man of real breeding [l'honnête homme].

When the École Freudienne was founded, Lacan envisioned what he called an "articulation with the mature sciences" [sciences affines]—that is, the sciences having an affinity with psychoanalysis. You did not have to be an analyst, or even be in analysis, to be made a member of the School; it could even be said that the more remote from psychoanalysis your primary affiliation, the greater your chances of being warmly welcomed. Those with philosophical training were especially well received, but mathematicians and lawyers even more so. This was a minor revolution, if you compare the success of this effort with the spirit of claustral, clublike jealously permeating the other psychoanalytic societies. Here was a discipline that no longer seemed to be afraid of contact with the outside world; in fact, it seemed to be actively courting such an encounter or, at any rate, a way out of an isolation that often meant enforced boredom. Secrecy was no longer a guarantee of strength, but instead simply revealed an underlying frailty.

Yet if the members of the École were so unafraid of the outside world, it was because they felt protected by the strength of their master, the man able to bring together people from all the sciences, yet who knew how to bring them very quickly into step with his own discipline. This openness to outsiders gradually changed into the conviction that, in this School, under the master's guidance, each discipline would be able to find the insight it still lacked. A sort of reversal took place: All these different forms of knowledge, whose questioning had initially been actively solicited, were now there for the sole purpose of validating psychoanalysis, and of convincing everyone of its superiority. A limitless pretention to being custodian of the truth, already so widespread among psychoanalysts of all persuasions, was laying the groundwork for an intellectual terrorism that would stifle all those who admitted their inability to understand, and who dared not to agree with everything that was being said and done.

I remember when Lacan was asked to discontinue his seminar at the École Normale, one of his students called to ask me whether I would agree to sign a petition to revoke the decision. I asked him to read me the text of the petition. His disciples used, among others, the following words to describe the room in which Lacan then taught (which to me seemed altogether fantastic, but unacceptable all the same): "The place where *science* is built." It was thus from 1968 on that an exorbitant belief in Lacan as the keeper of some great secret became widespread: He was the one who would be able to build, or rebuild, the unity of knowledge.

If we followed Lacan, it was because he was a prestidigitator of genius. He had begun by summoning all the disciplines to enlist their help in leading psychoanalysis out of its confinement. But once they had kept their appointment, the great doctor treated them like so many patients: He no longer saw in them anything but their wounds, flaws, and limitations. Since psychoanalysis had in the interim become an unrivaled specialist in wounds, flaws, and gaps, a specialist in defects of all kinds, it was as if, through being the custodian of defects, it had thereby become the discipline of disciplines–indeed, the science of sciences.

The same was true of psychoanalysis itself. After its initial stammering and blundering, it had at last found its fulfillment in Lacanian theory, and everything that had preceded it was relegated to a prehistory of ever-diminishing interest. Whence the conviction, still held by many, that, in order to master the analytic object and remain at the pinnacle of humanity, it is enough just to read Lacan, and never leave the confines of his writings. What good is it to go on asking questions that he has already answered for us, given that we don't have the time to devote ourselves to philosophy, mathematics, ethology, and Chinese in order to make sure that the whole thing actually holds together? How can one overlook the fact that, as reticent as certain people in our midst may have been, they were still, at one time or another, well and truly caught up in this comforting belief?

That a body of thought should be transformed into an institution is hardly exclusive to psychoanalysis. Nietzsche had seen it at work with the aging Auguste Comte;[1] and we now know that Heidegger was unable to avoid the same fate.[2] As for our contemporaries, they provide us with several instances of the same phenomenon. In a culture in which religion no longer plays its protective role, either the lesser minds are constrained to flock to a master thinker, in order to fill their persistent emptiness with his thought, or the great thinkers, having grown weary, are reduced to safeguarding their own precarious creations. Not only is Lacan no exception here, but he knew how to intensify the phenomenon by using the privileged means that psychoanalysis put at his disposal.

Transference has the particularity of raising its addressee to the level of a function. During the cure, and only during the cure, your analyst becomes the one and only, the incomparable analyst. But such a situation can be sustained only to the extent that an analyst is careful not to make his own particularities too obvious, and Lacan was brilliantly successful at appearing as the analyst in both public and private, or more subtly, acting in such a way as to make sure that all personal traits were concealed in constituting the figure of the analyst. Whereas Freud revealed himself through his dreams, his correspondence, or his case histories, Lacan did everything possible to avoid leaving any trace of his subjectivity. In order to safeguard the mystery of the analyst's position, and regardless of his company, he pushed this self-effacement to the limit. One of his longtime colleagues tells, for example, of one day being invited to share her thoughts on the Oedipus complex. She agreed, on condition that he in turn do likewise; but when she had finished her presentation, he disappeared, without offering any explanation. The relationship between the director of the School and one of its members could no more be reciprocal than could the relationship between analyst and analysand. Sometimes Lacan would also summon certain people to ask their opinion on some current matter, but he would refrain from offering his own opinion, even when questioned with some insistence. As for his discourse, it is well known that, after the fashion of analytic interpretation, it took on for his audience the obscure character of an oracle. Lacan wanted to remain the one who was supposed to know and, to that end, he set up an entire apparatus designed to prevent his knowledge from being put to the test.

To be sure, something of the order of transference—that is, a powerful relationship of trust and dependence—enters into every master–disciple relationship. But in other disciplines, there is nothing to be concerned about, since knowledge mediates the relationship. To the extent that a disciple has real access to it, it becomes possible for him to break away from the master's knowledge and to produce another form of knowledge for himself. By contrast, when disciples settle into their subordinate position, they can do little but repeat—and thereby veil—their own intellectual weaknesses. For his part, the master, wearied by the ongoing struggle with himself at the heart of his thought, turns this struggle into an institution, and could not fail to surround himself with mediocrities, for they are the only ones able to maintain in him the illusion of his constancy.

In psychoanalysis, the situation is different. If, as Lacan wanted, in the analysand's eyes, the analyst—through the play of the transference—does indeed become the subject-supposed-to-know (*le sujet supposé savoir*—he is undoubtedly something else altogether, but this is not the place to debate the issue), in reality, the analyst knows nothing, and it is up to the analysand to discover this for himself through the cure, draw the consequences, and make up his own mind. In adopting the position of master and producing pupils who were supposed to remain such indefinitely, Lacan was no longer merely *supposed* to know, he *knew,* and even saw himself as the only one who knew. Indeed, did he not repeatedly tell us that, on any given question, we could not go beyond the stage he had reached himself, and that to make further progress, we would just have to wait until he was ready to take another step? There was no longer any question of impeaching his knowledge, which he had transformed into an inviolable boundary. But, in doing so, he was diverting psychoanalysis from its goal, since we were becoming, in principle, incapable of breaking out of the role, no longer even of pupils, but of schoolchildren. Maintaining the transference was the condition of possibility of his mastery, which is why good disciples today still sneer at the possibility of resolving the transference.

Lacan also knew how to play on another characteristic feature of analysis. In the analytical cure, knowledge is illusory, in the sense that it stems solely from an assumption on the part of the analysand, and on which his discourse depends. Such knowledge intrinsically lacks substance, and exists only through the effects of a neurotic belief. One of Lacan's shrewd teaching practices (that is, outside the domain of the cure) was to systematically mimic the instability of analytic knowledge, thereby giving it an intrinsic stability. It was then no longer possible to do away with it, since the illusion of knowledge, which one could have denounced as analysand, was transformed on becoming an analyst into a knowledge that itself took on the form of the illusion. The absolutely contradictory relation between a stable knowledge open to criticism, and an unstable knowledge destined to fade away through the cure, took on the appearance of an intrinsic link. In what follows, we shall see just how difficult it is to disentangle this willfully obscure discourse, and why it has taken all this time to put one's finger on the contrivances.

Psychoanalysts cannot tolerate the irremediable opposition between their role as analysts, whose knowledge for the analysand can only be ephemeral, and what they say and write outside the analytical situation. They would naturally prefer their discourse to be derived from psychoanalysis as a science, whereas it is actually related to the form of infantile theories of sexuality. If Lacan was so successful, it was because he claimed to have found a remedy for the fatal contamination of analytical knowledge by the effects of practice. In Lacanian terms, it is an inescapable dilemma for every analyst: Either your knowledge is part of the rational order, in which case you are no longer an analyst, because you can no longer be assigned the function of the subject-supposed-to-know; or you are indeed an analyst, in which case your knowledge is tenable only in and through the assumptions of your analysands.

Thus, an absolute contradiction emerges between the Lacan who wanted to be the analyst *par excellence,* and the Lacan who assumed the position of the master who knew and taught what he knew. Either his knowledge was real, meaning that, by this very fact, he was no longer an analyst—that is, assigned solely the function of the "supposed-to-know"—or he really was an analyst, meaning that his knowledge was merely ephemeral, tenable only through the suppositions of his analysands.

Did Lacan really need student-analysands for his teachings to endure? At any rate, by confusing the two positions, he avoided all criticism, but at the same time he was bringing about the death of analysis. A typical Lacanian analyst, one of the fervently faithful, will today claim that should there be something he doesn't understand when reading Freud or Lacan, or if he has problems with the text, then he is always the one who is at fault. Freud and Lacan don't make mistakes, nor are they deceitful: A reinforced neurosis ends up supporting crass stupidity. All of this obviously reopens the question of the status of psychoanalytic knowledge, and it is understandable that, in order to fend it off, analysts insist that all critics should have undergone psychoanalysis, for they hope that the circle will remain vicious; in other words, that through the expedient of the transference and its lingering vestiges, each and every analysand will join the brotherhood of initiates and come to believe in its tenets.

Lacan saw the objection coming from a long way off, but rather than being discouraged by the problem, he simply systematized the common response. What would otherwise be seen as a defensive reflex action characteristic of a certain caste, Lacan sees as a theoretical necessity whose implications must be brought to bear at the institutional level. From 1948 on, he points out that, if it is to form the basis of a positive science, the psychoanalytic experience clearly has to be "universally verifiable" [contrôlable par tous] (*E*, 103). But it has to be clearly understood what "universally verifiable" involves here: An analysand's experience with an analyst can be resumed with a third party. That is, an analysand-become-analyst can pass on his knowledge to another analysand, who will himself become an analyst, and so on. We thus have an ongoing transmission, which, in multiplying analysts, tends to generalize itself, and which therefore can, in the end, become universal.

When he founded the École Freudienne in 1964, this was Lacan's sole aim. For example, in the "Preamble" published in the first yearbook, he writes: "Psychoanalysis currently has no surer way of commanding respect than by producing analysts." No doubt every society of psychoanalysts would acknowledge that one of its functions is to train analysts, but this function would always be subordinate to the therapeutic aims of analysis. Lacan subverts this hierarchy: For him, to quote from this same text, "pure psychoanalysis is not in itself a therapeutic technique;" pure psychoanalysis is didactic, that is, devoted to the production of analysts. This is why an analyst in the School becomes a didact, meaning a full-fledged analyst, not when he receives a certificate from the organization, but when he has transformed an analysand into an analyst. We can see, then, that the rather foolish objection raised by psychoanalysts who insist that all those having something to say about analysis should have undergone analysis has now taken on the dimensions of a first principle of analysis, and has in the process become the analytic institution's mode of functioning. If you want analytical knowledge to be generally recognized, it is both necessary and sufficient to produce analysts who will in turn produce others, and the population of analysts will then tend asymptotically to overlap with the population *tout court*. Once there is no longer anything *outside* psychoanalysis, its teachings will no longer meet with any objections and will be considered valid. The "initiatory path" [la voie initiatique] will then truly be transformed into "transmission through recurrence" (*E*, 103).

This is why Lacan, with the same intrepid logic, far from regarding the transference as an obstacle to the transparency and objectivity of knowledge, introduced it as a necessity into the field of this same knowledge. In an appended "Note," also published in the School's first yearbook, it is stated that "the teaching of psychoanalysis can only be conveyed from one subject to another by means of transference-work [le travail de transfert]."[3] This is a strange notion, since transference no longer refers, as it does in Freud, to the motor of the cure, but becomes instead the condition for a form of knowledge. Here again we find the confusion noted earlier, but this time it is explicit and quite deliberate. Whereas in the course of the cure, transference is supposed to fade away once the analyst's knowledge has been repudiated, the transference-work now becomes a way for the analysand to participate in his analyst's knowledge (or that of the analytic society to which he belongs); and, since the transference is tied to an ongoing effort [un travail], the process can continue indefinitely—no more transference, no more transference-work [travail], and thus no more knowledge. Should one leave the sanctuary of the transference, however, the only option is to succumb to theoretical drift. Did Lacan not repeatedly say that those who left him could only end up in sterility and ignorance? For it is obvious that the transference-work did not refer to just any analyst, but to the director of the School himself. And the "teaching" in question was his, a teaching that reverberated in and through the students he had shaped.

It should be noted in passing that, even if this notion of transference-work should seem an aberration in the eyes of a Freudian analyst, it is, on the one hand, perfectly in keeping with Lacan's aim. In his seminar on the transference, he spoke of entering the said transference in the register of the desire for knowledge. On the other hand, all psychoanalytic societies are faced with the same problem through the teaching they dispense, and being unaware of the problem is no guarantee against falling into a Lacanian rut.

The *pass,* too, was invented in order to focus the analytic institution on the production of analysts. It involves deploying an apparatus designed to isolate the moment at which the analysand becomes an analyst, with a view to eliciting its characteristic features, and allowing a procedure tailored to this end to produce its theorization. The pass was supposed to be a kind of laboratory in which the analysand becoming an analyst would be the object of the experiment—which was, after all, only logical. Since the institution was created to produce analysts, it was important to acquire a theoretical knowledge of the emergence of the phenomenon, so that the production (the word is not idly chosen) would be the result not of an uncontrollable proliferation, but of a rule-governed procedure, and thus attain scientific status. If Lacan invented this apparatus and carefully supervised its functioning, it was probably because he expected it would provide a means of systematizing what was in fact the nerve center of his enterprise, namely, the secret of producing analysts, which was gradually going to alter social relations by universalizing the adoption of analysis. In short, knowledge of this moment of passage was the key to "transmission through recurrence." But it should come as no surprise that Lacan was completely disappointed by the results of his experiment, since the procedure simply multiplied the contradictions mentioned earlier. How could knowledge of becoming-an-analyst be attained if analysis itself was reduced to a form of knowledge (albeit knowledge of desire), whereas this "becoming" is intrinsically linked to the necessity of pursuing one's own analysis, that is, to the necessity of continuing to cure oneself in the process of curing others? Recognizing that becoming-an-analyst amounted to this would have reduced to nothing the novelty of having reversed the aims of analysis, which meant you had to turn a deaf ear whenever you heard what was actually being said about the pass. Moreover, given that it is founded on the most inalienable personal responsibility, how could the pass be transposed onto the institutional scene, a scene exclusively constituted by relations and effects of power? Whatever the criticisms leveled at it, it has to be recognized that, among other inventions designed to create an analytic institution responding to the specificity of analysis, the pass had the power to enthrall us and to hold us in its snare. There were some who immediately sensed its inherent perversion, but there were many others whose eyes were opened only after its devastating effects had been felt and confirmed.

Lacan did not content himself with theorizing analysis as the production of analysts. He did everything he could to implement it socially, first by rejecting the rigid standards imposed by the International Society. For neither the obstacle course presented to the candidate's flexibility nor the competency regulations guarantee the quality of an analyst; they lead, rather, to conformity. And second, Lacan pursued on behalf of his School a systematic policy of placing his disciples in Centers of psychiatry and psychotherapy, which enlarged the sphere of contact with potential clients. But above all, Lacan produced analysts, because his personality and discourse attracted a multitude of applicants with very diverse backgrounds. He was, in a sense, the prestigious showcase of a great enterprise. Because treatment in analysis was no longer centered on the cure, but now seemed to be a privileged site of introduction to the secrets of culture, there was no longer anything shameful about being psychoanalyzed. To become an analyst recognized by Lacan meant acquiring a kind of nobililty, which in turn opened up the possibility of acquiring a clientele. Without him, a number of us would never have attained such status and would not have had the means to make a living from it. After having given existence to many, analysis also became, thanks to Lacan, the means of their subsistence.

Finally, one of the most powerful elements in Lacan's seductiveness lay in the fact that he pushed the contradictions and paradoxes of psychoanalysis to the limit. In his research, the more he encountered impossibilities and impasses, the more he claimed to make them the very cornerstones of his system. This exorbitance, together with a keen sense of the tragic, made him irresistible for French intellectuals:

> Lacan is steeped in Greek hubris, impossible exorbitance and deadlock. The Greek hero, the tragic hero, of which Lacan is a perfect stylistic and theoretical model, is situated beyond all forms of distance. Where other men, those who live in the city find, as best they can, ways of adapting an empty space, the tragic hero crosses the frontier and plunges headlong into monstrosity: he is at once God and animal. Incestuous, or enraged, possessed, he is no longer a "man": he has lost all sense of human standards. He has forgotten the lesson of myths, which are, as Levi-Strauss tells us, lessons in "appropriate distance" [bonne distance]. Keeping the right distance between yourself and the madness of an impossible desire, between yourself and the real: but this distance actually exists—it is regulated on all sides by the multiple codes of so-called everyday life. As described by Lacan, psychoanalytic practice consists, on the contrary, in *exacerbating distance*. Let there be no misunderstanding, you wretched souls: your desire is forever cut off from its object, which is lost, and will always be undermined in the most agonizing separations. And then there are those sublime sentences of his, which catch the intellectual's ear for the tragic, the intellectual who is always willing to allow himself to be seduced whenever the impossible is proposed as such; those sentences, whose lulling rhythm panders to a delight in the loss of the lost paradise.[4]

This was a figure who seemed tragic in his moment of glory, but perhaps even more so in his decline—for example, when in June 1979 he concluded the Congress on the transmission of psychoanalysis. After having evoked "this admittedly fairly wild story called the unconscious, and which is quite possibly a Freudian delirium," he comes to the subject of the pass, saying that in it he had "expressed his faith in something called the transmission of psychoanalysis, if there were such a thing as transmission." For he immediately goes on to say: "In the form in which I currently conceive it, psychoanalysis is not transmissible. It is tiresome indeed that each analyst is forced—since he certainly has to be forced into it—to reinvent psychoanalysis." He then raises the question of the cure: "It is a fact that there are people who are cured. Freud certainly made it clear that an analyst need not be possessed by the desire to cure; yet it is a fact that there are people who are cured, cured of their neuroses, indeed, cured of their perversions. How is this possible? In spite of everything I have had occasion to say on the subject, I just don't know." In the twilight of a life devoted to laying new foundations for psychoanalysis, such admissions open an unplumbable abyss in the completed work and can hardly fail to have an effect on a retrospective reading of it.

In what follows, I do not intend to analyze Lacan's entire *oeuvre;* my intention is rather to go over some essential points in his doctrine in order to reveal its basis and show how it progresses. To do so, the texts selected will need to be followed step by step. What actually counts is the movement of his thought, since the detours are just as important as the conclusions. Lacan is a strategist who does not make a move without precise intentions, or again, he is a rhetorician out to persuade. It is therefore important at each stage to ask yourself what his point of departure is, and where he wants his reader or audience to end up. Difficulty in understanding him most often comes from being overeager to derive some kind of profit from his assertions, or from habitually succumbing to the play of associations prompted by what one reads or hears. Lacan is actually quite explicit; and yet you do have to know how to read him, and not take his meanderings, which are deliberately designed to win your allegiance, for useless excurses.

I have used only texts which have already been published in French, so the reader may refer to them if he sees fit. The trouble with citations from the relatively inaccessible seminars is that they are stripped of their context, and only with some difficulty can a given interpretation be contested. Given that my purpose here is not to account for Lacan's entire *oeuvre,* but to bring out the style or, better, the turns of his thought, it is hardly necessary to analyze everything, since an author's particular style can be detected everywhere. I have, however, chosen texts from different periods, in order to determine whether Lacan's thought changed over the course of time, but also because, in posing the question of style, there is a further question to which I should like to provide an answer: Does the system, or doctrine, or theory proposed by Lacan hold up when subjected to a minimum of logic and rigor, a criterion that an intellectual enterprise seeing itself as a science can scarcely avoid?[5]

Notes

1. *Daybreak,* trans. R.J. Hollingdale, Cambridge, C.U.P., 1982, pp. 214–16.
2. The phenomenon must have begun at a very early stage; cf. Elisabeth Young-Bruehl, *Hannah Arendt,* Editions Anthropos, 1986, p. 75, 397–98, 583.
3. I have translated the phrase "le travail de transfert" as "transference-work," on the analogy of the Freudian notion of "the dream-work;" but "le travail de transfert" is to be understood as functioning *outside,* and as *subsequent to,* the analysis proper. [Tr.]
4. Catherine Clément, "Les allumettes et la musique," in *L'Arc, D.W. Winnicot,* no. 69, p. 66–67.
5. I should like to mention several authors whose work initially suggested this form of reading to me: Jean-Luc Nancy, Philippe Lacoue-Labarthe, *Le titre de la lettre,* Galilee, 1973; Regnier Pirard, "Si l'inconscient est structuré comme un langage," in *Revue philosophique de Louvain,* Nov. 1979; Phillippe Julien, *Le retour à Freud de Jacques Lacan,* Erès 1985, a work which gives dates for all citations, and brings out Lacan's repeated reworkings; Marcelle Marini, *Lacan,* Pierre Belfond, 1986—a masterful book, henceforth indispensable for any reading of Lacan; Monique David-Ménard, "Transfert et Savoir," in *Esquisses psychanalytiques,* no. 4, autumn 1985, an article that outlines in detail the movement of Lacan's thought in the seminar on the transference.

2

The Science of the Real

The Ambiguities of Science

Lacan's *oeuvre* has the appearance of one of those trees found in India, whose branches can go on to become just as many trunks, in an extension that seems as if it must go on forever, unstoppable in its unwavering appetite for space. Any one of the subjects he dealt with in his seminars could be privileged in being chosen as the point of departure for recomposing his doctrine in its entirety. Moreover, of all the questions concerning those he would call "speaking beings" [les êtres parlants], there was not a single one that did not at one time or another steal or smash its way into the scene he was performing before us. This is a discourse, then, in which the center is everywhere and each and every element can be made central.

The form of his discourse was itself also subject to this double impression of being at once a rigorous sequence and an uninterrupted digression. Listening to him, you never knew whether to place what he was saying under the sign of logic or to regard it as the most extravagant fable. Most of the time he would improvise on an outline prepared in advance, occasionally letting himself drift along detours provoked by some chance detail: A remark by the previous speaker would suffice, or a book just barely off the presses, or maybe the arrangement of the room in which he was officiating. But he did none of this without a definite intention, always holding firmly to a single guiding thread, which an attentive reading will subsequently reveal.

As for his published work, it would have been nice had it been contained in a well-ordered series of books, with a uniform format. But the publication of the seminars has been, at best, parsimonious, yet withholding them has simply given rise to their systematic proliferation as unwieldy photocopies in all shapes and sizes—along with a superabundance of bad readings. These processes only seem to better serve the dissemination of a secret doctrine bent on overrunning everything in its path, but they also preclude one from having an overview of a body of work to which death has, however, now put an end. There is always an other Lacan, always other texts where he said something else, other discussions which re-pose questions to which you thought you had the answers, other confidential papers that could well revive old controversies.

How to grasp such an *oeuvre*? And what approach should be taken, given that, with its highly "written" form, Lacan's style is wrought of tight juxtapositions and dazzling insights that whisk out of sight the very things they were supposed to elucidate, of contradictory aphorisms that impose risky, always debatable interpretations. His style, then, is specifically designed to ward off simplification, but more generally, to defy any translation that would gloss over the words at issue and in need of explanation, words that should therefore be left to their diamondlike indivisibility or, in the final analysis, their foliagelike obscurity.

Yet if we are to escape the obscurantism which Lacan himself detested, we must certainly make an effort to understand his work, *resolve* to understand it, and thus convey something of it outside the sect that insists on unquestioning acceptance of both the doctrine and the ponderous, jargon-ridden language [langue de bois] in which it is couched.

When it comes to clarification, all points of view are in some sense arbitrary; yet there *is* one that can be privileged, since it will allow us to furrow a way through the entire *oeuvre*, grasping its diversity while respecting its unity. It was the project Lacan pursued relentlessly: To turn psychoanalysis into a science. To be sure, the project was progressively modified, and its various stages must, of course, be delineated; and using this project as our guiding thread is precisely what will make it possible to distinguish his successive attempts as so many formulations of, and provisional solutions to, a single problem. Rather than "project," though, it would perhaps be better to talk in terms of a spectral presence, for there is a considerable—and noticeable—distance between what Lacan constantly dreams of achieving and what he actually presents as a result or recognizes as an inevitable and even felicitous setback.

Such an approach to Lacan's *oeuvre* offers the further possibility of appropriately situating his adversaries and, through all the polemicizing, of rendering intelligible his doggedness and insistence, indeed, his unilateral or overly hasty assertions. It is precisely because he is carried away by the necessity of his project that Lacan comes to hurl challenges at himself, challenges that, outside this tightly closed field, would become totally unintelligible, so much do they seem to be driven by an utterly boundless ambition.

Thus, in 1936, we find a capitalized subheading boldly asserting that "psychology is constituted as a science when Freud posits the relativity of its object, while still restricting it to the facts of desire" (*E*, 73). The words "restricted relativity," making Freud nothing less than an emulator of Einstein, here make it plain that physical science is already serving as a model. I say "already" because mathematical physics will in fact never cease to be the model *par excellence* for Lacan, and because the explicit effort to align psychology and, before long, psychoanalysis with this model, will reappear later on numerous—and decisive—occasions. Even if the proposed solutions are themselves subject to variation, the model itself undergoes not the slightest retouching.

What strategy does Lacan adopt in "Beyond the 'Reality Principle' " in order to give the impression that psychology can henceforth be considered as having entered the domain reserved for science? He begins with a critique of associationism: It is neither positive nor objective and, rather than basing itself on the constitution of form (Gestalt) and on phenomenological analysis, it is founded on mythical notions and the search for what is truly real. It is a question here of avoiding "scientism" by opposing to it a conception of science that "can honor its ties with truth," but which "can in no way take truth to be its own goal" (*E*, 79). The text continues:

> If this seems like some kind of trickery, then pause for just a moment over the actual criteria of truth, and ask yourself what remains, within the vertiginous relativisms of contemporary physics and mathematics, of even the most concrete of these criteria: what has become of *simple conviction,* the test of all mystical knowledge; of *self-evidence,* the very basis of philosophical speculation, and even *non-contradiction,* that most modest requirement of the empirico-rationalist edifice? But more within the scope of our own judgment, can it be said that the scientist asks himself whether a rainbow, for example, is *real?* What matters to him is that a phenomenon is communicable in a certain language (a condition of the *mental order*), able to be recorded in a certain form (a condition of the *experimental order*), and able to be successfully inserted into the chain of symbolic identifications in which its particular science unites the various aspects of its object (a condition of the *rational order*) (*E*, 79).

For now, let us emphasize, in order to come back to it shortly, that in this definition of mathematical physics, there is curiously no mention of its concern with quantification, nor of its quest for verification, and equally scant mention of its constant use of deduction grounded in the identity principle—in other words, the principle of noncontradiction. For psychology to become a science comparable to physics, one must therefore start out by reducing the scientificity of the latter to terms that are either sufficiently vague or willfully equivocal.

Next, the specificity of psychical reality has to be brought out by describing the analytical experience in which "this posture in Freud of submission to the real" (*E*, 81) manifests itself. In the analysand's discourse, this experience will tend to isolate both what he does not understand and what he denies, the image forming and informing him, an image that is indeed the "real" to be sought beyond the disparate reality of thoughts and feelings. This experience will be called "scientific" because, in the first place, it is referred to the domain in which the scientist makes discoveries whose geneses are passed over in silence, so as to give the impression that "his methods conform more closely to the purity of the idea" (*E*, 86). It is equally scientific inasmuch as physical science is subject "in all of its procedures to the form of mental identification" (*E*, 87); in other words, theories remain the expression of a form which is "constitutive of human knowledge" (*E*, 87).

From there, Lacan goes on to deduce that both the object and method of psychology are not subjective, but relativist (a further allusion to Einstein), because they are founded on interhuman relations. And these relations are constituted by identifying with counterparts [semblables] transformed into images that both inform the subject and determine the ways he lives and thinks.

Then comes the final step: Psychology is said to be a science because it is based on the energetic concept of libido, which is "merely the symbolic notation for the equivalence between the dynamisms that such images invest in behavior" (*E*, 91). It can then be concluded that:

> Through this notation, the efficacy of such images, without yet being amenable to a single form of measurement, but already equipped with either a plus or minus sign, can be expressed through the equilibrium that they set up, and by employing, as it were, a method of *double weighing* [double pesée]. As used here, the notion of libido is no longer metapsychological: it is the instrument of psychology's progress towards a form of positive knowledge [savoir]. For example, combining this notion of libidinal investment with a structure as concretely defined as the super-ego represents—over the idealizing definition of moral conscience, as well as the functional abstraction of those reactions known as *opposition* or *imitation*—progress of a kind comparable only to the introduction into the physical sciences of the *weight over volume* ratio, when it replaced the quantitative categories of heavy and light. The rudiments of a *positive* determination have thus been introduced into psychical realities which a relativist definition has made it possible to objectify. This determination is dynamic, or relative to the *facts of desire* (E, 91).

For such an outcome to make sense—that is, for it to be established that psychology is now walking the path of science, what are the necessary steps?

1. In order to avoid being accused of proposing to turn psychology into a second-rate science, Lacan provides himself with the most incon-

testable model, the one that, through its access to mathematics, has rid itself of any dependence on the qualitative. But for the model to be usable in this instance, its overly sharp edges have to be rounded off. Its universality, based on the fact that it needs only certain kinds of knowledge and no interpretation to be understood, is reduced to communicability in "some form of language." But mathematical language is not just *any* language—it is highly specific. Its relation to experimentation, verification, and refutation—all essential, even if there is some time lag (the Sputnik was the first experiment with Newton's laws)—is reduced to the simple possibility of being recorded. Many things are recorded in meteorology, yet as far as we know, it is not dependent on mathematical physics. And as for physics' ability to express its theories and results in algebraic formulae, this supposedly amounts to nothing more than inserting a phenomenon "into the chain of symbolic identifications in which its particular science unifies the diverse aspects of its object," an elegant formulation in which the words "symbolic identifications" prepare us for the slippages of meaning to follow, but in which the formal rigidity of sigla stripped of any evocative potential is lost. It is true—and not without a degree of cunning—that Lacan chooses the rainbow's appearance in the sky as his example of a physical phenomenon, an example lending itself more to poetic *élan* than to the austerity of a calculus.

Lacan's first procedure can be summed up as follows: After having selected the very best model of science, its most prestigious traits had to be somewhat downgraded, and others selected, equivocal enough to be applied indifferently to both science and psychology.

2. The second procedure consists of calling on science to dredge up its origins: There *are* no discoveries that might not have made their way through dreams, distractions, nocturnal anguish, deeply-rooted errors, or extravagant beliefs—in short, through what is most irrepressibly subjective. But unlike art, science precisely *must* efface such moments from its genesis in order to subject itself to the strictest rigor and rationality, stripped not only of all emotion, but of any trace that might refer it to a specific person. If psychology cannot detach itself from "personal commitment" (excluded by science), since this is its very object, one would normally be forced to conclude that psychol-

ogy is not, and, granting this, *cannot* be an exact science. But Lacan draws just the opposite conclusion, scoffing at any science that presumes, "like Caesar's wife, to be above suspicion" (*E*, 87). He introduces this confusion in order to support the following reasoning, which we shall encounter again later: Science excludes subjectivity, but it should not do so, since subjectivity is a necessary part of its constitution; therefore, the discipline that concerns itself with subjectivity can be deemed scientific. Or, put more succinctly, what science excludes becomes, by its very exclusion, scientific. The confusion was thus an essential part of the proof.

3. A similar procedure is then used to remind physical science of the anthropomorphism of all knowledge, of which it forms a part. Lacan invokes Emile Meyerson, who is said to have demonstrated that the structure of intelligence is "subject in all of its procedures to the form of mental identification" (*E*, 87). Lacan neither quotes nor refers directly to Meyerson's works, for the very good reason that Meyerson does not *use* the expression "mental identification," although it does correctly "translate" his thought.[1] Against scientism, his aim is rather to bring out an analogy between unconscious processes of knowledge as conceived by "common sense," and the conscious processes of scientific thought.[2] Lacan reasons, implicitly, in the following way: If physics, as a science, is based on a belief in the similarity of nature and thought, then psychology, being based on "affective communication, which is essential to social groups" (*E*, 87), is itself a science. It is apparently even *more* scientific than physics: "Indeed, man maintains ties with nature which involve, on the one hand, properties of a mode of thought characterized by identification, and on the other hand, the usage of instrumentation or artifical implements. Man's relations with his counterparts (semblables) take a much more direct route" (*E*, 87).

Now, not only does Meyerson not mention identification, in the sense of imitation or introjection (a necessary median term between physics and psychology), but he aligns identity with immutability, variability and constancy, and so on, which have nothing to do with the alterity requisite to psychology. Moreover, the epistemologist, who asks himself where the principles and theories of physics come from, must not be confused with the scientist himself, who uses these same principles and theories to perform experiments and produce equations. Thus, even if Meyerson *had* spoken the same language as Lacanian psychology, the links with science proper would still need to be established.

4. Reading Meyerson's *La déduction relativiste* gave Lacan the idea of using this same adjective in his paper to designate interhuman relations. Meyerson obviously uses the word *relativist* because he is dealing with Einsteinian theory, one of the goals of which is, as we know, to eliminate the observer more completely from the field of physics, a goal which therefore happens to be diametrically opposed to the project for a psychology as expounded by Lacan. But the word is not used innocently, of course; it is meant to at least perturb the reader and have him assume that the newly expounded psychology is inscribed in a similar domain and has thus attained a comparable degree of scientificity.

5. Lacan's final twist in this text consists in treating the libido as a symbolic notation, which "without yet being [sans pouvoir encore] referable to a single unit of measurement," marks "progress of a kind comparable only to that brought to the physical sciences by the use of the weight over volume ratio" (*E*, 91). Everything hinges on the "not yet" [pas encore]: Thanks to this phrase, what he actually says is true, but the text achieves its effect on a well-disposed reader nonetheless. From what Lacan is suggesting here, it can be inferred that psychology will come to master quantity and make it a pivotal part of its construction, and all the more so when a probable slip of the pen begs the question and imposes this interpretation, since what Lacan actually writes is that the weight over volume ratio "has been substituted for the quantitative (instead of 'qualitative') categories of heavy and light" (*E*, 91). Thus, even when physics was still dealing with the heavy and the light, it was already in the domain of the quantitative—and likewise, *a fortiori,* a psychology which speaks of libido. Yet, if there is a mythical concept *par excellence,* the libido is surely it, and Lacan will by no means forego denouncing it at a later stage.

The entire strategy of this text can be summed up by pointing out Lacan's deliberately equivocal use of three keys words: *identification, symbolic,* and *relativist.* Identification is said to subtend man's relation to nature and to define physical science, but it is simultaneously seen as the phenomenon Freud described whereby one person models himself on another in one or more respects. As an adjective, symbolic refers either to the algebra used in physics, or to language in analytic practice. And finally, relativist is an allusion to Einsteinian theory, while in other respects it covers interhuman relations. Only by conceding an equivalence between these terms could the claim for the scientificity of psychoanalysis be considered valid. Since they are not equivalent, we are forced to conclude that, instead of witnessing the exposition of a coherent and well-founded body of thought, we are in the presence of a surrealist collage making it possible for psychology to dream of better days ahead.

The interest of this by no means negligible text lies in its description of the analytic process as the possibility of isolating the image that forms and determines behavior. But this is obviously only a description, and—how could one fail to notice?—only one description among others that seek to bring out the specificity of the analytical experience. You have to give Lacan credit for having seen very clearly which conjurer's tricks were needed in his attempt to bestow scientific status on psychoanalysis, for if he hadn't, he would not have tried other methods in order to discover true foundations. For him to have ceaselessly imagined new solutions, right up until the end of his life, each one of them doomed to failure, he must certainly have been driven by a necessity that cannot simply be a reflection of his personal ambition. Without going into it here, since this is not the appropriate place, it should be noted that, if psychoanalysts did not see themselves as walking the scientific path, they would be forced to recognize that as a practice, psychoanalysis has more to do with art or perhaps even with ancestral healing techniques, and it would lose part of the prestige and authority accorded to it, which analysts feel has been legitimately acquired.

For the time being, though, and for more than ten years (from 1936 on), Lacan will remain faithful to the same perspective, as evidenced by his 1948 report on "Aggressivity in Psychoanalysis." In this text, he says he wants "to develop a concept capable of scientific usage, that is to say, capable of objectifying facts of a comparable order in reality, or, more categorically, of establishing a dimension of experience whose objectified facts might be regarded as variables" (*E*, 101). To be sure, he does acknowledge that analytical experience is essentially linked to subjectivity, but this is simply an opportunity for him to return to his model: "It cannot be objected that, from the standpoint of the ideal standards met by physics, this subjectivity must necessarily be ephemeral: physics does eliminate subjectivity by means of the recording apparatus, but it is not guaranteed against personal error in the reading of the result" (*E*, 102). Since, in analysis, the "subject presents himself as capable of being understood" (*E*, 102), the comparison is justifiable. It is even possible to deflect one final objection: "Can its results form the basis of a positive science? Yes, if the experience is universally verifiable" (*E*, 103). This is what he means by "universally verifiable" (contrôlable par tous):

> This experience, constituted between two subjects one of whom plays in the dialogue the role of ideal impersonality (a point which will require our attention later), may, once it is completed, and provided that if fulfills the conditions of competency required of any special research, be resumed by the other subject with a third subject. This apparently initiatory path is merely transmission by recurrence, which should surprise no one, since it stems from the very bipolar structure of all subjectivity. Only the speed of diffusion of the experience is affected by it, and even if its restriction to a particular cultural area may be debatable, although no healthy anthropology could find fault with this, everything would indicate that its results may be sufficiently relativized to provide a generalization capable of satisfying the humanitarian postulate inseparable from the spirit of science (*E*, 103).

This clearly means that for the experience in question to be universally verifiable, it is sufficient for everyone to have themselves psychoanalysed. Argumentation like this leaves you truly dumbfounded. Einsteinian physics proposed to eliminate the observer, but by *removing* him from the field of investigation, by reducing to zero the part played by subjectivity. Lacan too eliminates the observer, but by absorbing him into the system, by handing it over entirely to subjectivity—in other words, by annihilating all possible objectivity on the part of the observer. In this passage, then, the general and the universal—or, if you prefer, the infinite and the indefinite—are merged and confused, meaning that if a single person happened not to be analysed, the entire argument would collapse. It will be objected that Lacan did not stop there, that he went on to abandon this conception of psychoanalysis as a dual relation. Here, however, it is not the content of his thought that matters, but its style, which is essentially based on the construction of equivocation; from this style, he never deviates.

The developments proposed in the text "Aggressivity in Psychoanalysis" are not without interest, but once again, they pertain to the register of description and "phenomenological analysis" (*E,* 76), and as brilliant as it may be, such a description can never be seen as belonging in the domain of science.

The notion of *symbolic,* already present in the texts cited above, becomes more explicit in the paper presented in May, 1950, "Introduction théorique aux fonctions de la psychanalyse en criminologie." Here he invokes Mauss:

> Let us, then, resurrect the limpid formulations that Mauss's death has brought to our attention; social structures are symbolic; to the extent that he is normal, the individual uses these structures in actual conduct; to the extent that he is psychopathological, the individual expresses them through symbolic forms of behavior (*E,* 132).

Whereas the link between symptoms and the word *symbolic,* used adjectivally, had been pointed out earlier (*E,* 103), here Mauss's work compels Lacan to describe the pathological as a function of the environment.

But other conceptual links come to be established, no doubt due, as before, to the influence of ethnology:

> If the concrete reality of crime cannot even be grasped without referring it to a symbolism whose positive forms are coordinated in society, but which is inscribed in the root structures unconsciously transmitted by language, this symbolism is also the first in which psychoanalytic experience, through pathogenic effects, has demonstrated to just what previously unknown extent this symbolism reverberates in the individual, in his physiology as well as his behaviour (*E,* 129).

Here the move is made from society to the individual, with the help of three words that are in for a very promising future: Symbolism, structure, and language; but Lacan does not yet see how to make use of them in the framework that continues to haunt him.

It is Levi-Strauss, with his *Introduction à l'oeuvre de Marcel Mauss,* which appeared in 1950,[3] who will open the way to a complete renovation.[4] Although Lacan does not mention this text, and never cites it explicitly, for various reasons it is certain that he read it. First, he always kept abreast of the works of Levi-Strauss, as the numerous references to them in the *Écrits* testify; second, a passage from Levi-Strauss's "Introduction" is quoted twice, once in 1953 and again in 1960 (*E,* 279; 821). Finally, in a note in this same text, Levi-Strauss refers to "the profound study by Dr. Jacques Lacan, 'Aggressivity in Psychoanalysis'." This merits some attention.

In his text, Levi-Strauss writes: "Properly speaking, it is the one we call 'sound in mind' who is alienated, since he agrees to live in a world definable solely by the relation of self and other" (p. XX), a remark that he comments on in a note: "Such is indeed the conclusion drawn in the profound study by Dr. Jacques Lacan." Lacan actually concluded nothing of the kind: He contented himself with describing the forms and aspects of the relation termed "Imaginary." In citing Mauss, he was undoubtedly pursuing his dialogue with the ethnologists, but Lacan would not have endorsed what Levi-Strauss goes on to say: "The health of the individual mind implies participation in social life, just as the refusal to do so (a refusal which itself still follows prescribed forms) coincides with the emergence of mental disturbance." Lacan would never say such a thing, for that would mean being reimmersed in the horrors of social adaptation and—who knows?—the much-scorned "American way of life." But what matters here is that Lacan is strongly influenced by a man who is well-situated and who acknowledges him, a man he will hold in high esteem and on whose authority he will rely for some 15 years.[5]

Why insist so much on this influence? Quite simply because these forty pages from Levi-Strauss encapsulate everything that Lacan was waiting for in order to reactivate his project to make psychoanalysis a science.[6] Indeed, Levi-Strauss proposed to resolve the problem of the scientificity of ethnology by appealing to the unconscious[7] as simultaneously participating in the subjective and the objective—an unconscious amenable to being treated, in the manner of structural linguistics, as a language whose combinatory would need to be made explicit. Levi-Strauss goes even further; reducing the mana to a zero-degree symbol (this is the passage Lacan refers to: *E*, 279, 821), he rids ethnology of everything pertaining to the sacred and the mysterious—in short, of everything that would by definition elude science.

As if brought to him on a platter, Lacan now had everything that, until then, he had lacked—and he certainly put it to good use. It takes him 3 years to assimilate his new source of inspiration, but it can be said that everything that is central to the 1953 "Discourse of Rome," what gives it its novelty, descends directly from the *Introduction à l'oeuvre de Marcel Mauss*. Lacan will even go so far as to adopt its extremely precious equivocations.

As an ethnologist, Levi-Strauss undoubtedly remains in the domain of the social and the collective. Nevertheless, he largely opens the way for Lacan's borrowings when he asserts that "Like language, the social *is* an autonomous reality (in fact, the same reality); symbols are more real than what they symbolize, the signifier precedes and determines the signified" (p. XXXII). He thereby posits the autonomy of what Lacan will single out and substantify as the Symbolic. Levi-Strauss is then inspired by structural linguistics to mathematicize "the unconscious mental structures accessible through institutions and, better still, in language" (p. XXXIX). Institutions then need only to be replaced by Oedipus and the Law. Finally, through this mathematicization of language that reveals the unconscious, Levi-Strauss feels he can exclude everything pertaining to "feelings, wishes and beliefs" (p. XIX), which will make it possible to do the same in psychoanalysis—that is, to no longer bother with whatever has to do with the affections, the imagination, lived experience, the unspeakable, and the unfathomable.

This whole fine edifice can hold up only if one maintains the identity of two irreconcilable definitions of the Symbolic: On the one hand, the one that inaugurates exchange and is tied to language and meaning, and on the other hand, the one that refers to the algebraic sign and, by definition, cannot and must not signify anything. Not only does Lacan not resolve this ambiguity, an ambiguity he had already exploited (*E*, 79), but he makes systematic use of it when speaking of his "algebra" (which is never anything more than a figure of speech), claiming that it accounts for actual discourse.

At the end of his *Introduction,* Levi-Strauss presents a further notion that Lacan will also appropriate: The notion of the floating signifier, the excess of the signifier not yet used in order to signify, to which he will attribute a symbolic value of zero, "a symbol in its pure state, thus capable of taking on any symbolic content whatsoever" (p. XLIX). This conception is clearly a stage on the way to the autonomy of the signifier with respect to the signified.

If one now rereads the second part of the "Discourse of Rome," a decisive text for the scientific fate of psychoanalysis, exactly the same ingredients are found: The symbol, language, the unconscious, mathematicization through linguistics, and the pathological. It is enough to keep Levi-Strauss's text in mind to notice that constant reference is made to it; but because its field of application is no longer ethnology, there are going to be some substantial distortions.

To begin with, the symbolic order can no longer be related to the social, meaning that language alone is now going to bear its entire weight. To speak of a symbolic order for individuals considered in isolation is nonsense, since the word "symbolic" is meant to evoke exchange; so the Symbolic is then merely the ensemble of words that make up a language. Even if Lacan does allude to social facts, he nonetheless separates the symbolic order from the social group which is at once its effect and its support. This is why the term will have to be substantified and substantialized, turned into a substantive (whereas it had previously been an adjective), and into a substance, for, since it is no longer supported by anything else, it must now support itself. You could readily concede that the social precedes the individual, that every human individual, entering a determinate society at birth (or, if you prefer, even before birth), should be marked by its functioning and should have to submit to it on pain of exclusion or madness. But is not so easy to understand how it can be claimed that "Freud's discovery is that of the field of incidences, in man's nature, of his relations to the symbolic order" (*E,* 275), if "the law of man is the law of language" (*E,* 272), if the symbol is fully realized on becoming language, and if "freed from its usage, the symbolic object becomes the word" (*E,*

276). For, in that case, in order to produce a man, it would suffice for him to learn to talk to a machine, any relation with other human beings having become superfluous. But, even if he is called a "speaking being," what kind of a man would this be since language alone would still not impel him into the exchange implied by a given symbolic system, of which language is only one element.

In abandoning the social, without which the Symbolic has no support, Lacan is thus forced to substantify speech, to give it a certain power (*E*, 279). He is also forced to substantify language, and claim that "the concept, preserving the duration of the fleeting [sauvant la durée de ce qui passe], engenders the thing" and that it is "the world of words which creates the world of things" (*E*, 276); in short he is forced to reestablish the theology of Creation through the Word. Well aware of what he is doing, Lacan knows perfectly well that an implacable logic is leading him in this direction. Psychoanalysis needs a theological foundation in order to sustain itself: It is not by chance that Lacan cites the Gospel according to St. John at this point in his discourse, nor is it accidental when, further on, he reworks Goethe's expression. But if Lacan wants us to believe that he is still following Levi-Strauss here, the latter is now no longer following Lacan. The following year, Lacan refers to some private conversations with Levi-Strauss:

> Levi Strauss is beginning to back away from the very sharp partition he set up between nature and the symbol, a partition whose creative value he nevertheless still feels, since it is a method which allows you to distinguish between registers and, by the same token, between orders of facts. He is wavering, and for a reason which to you may seem surprising, but which he freely admits—he fears the re-emergence of a kind of transcendance in the guise of the symbolic register, for which, in his affinities and personal sensibilities, he feels only dread and aversion. In other words, he fears that, after having expelled God through one door, we might be ushering him back in through another. He does not want the symbol, even in the extraordinarily refined form in which he himself presents it to us, to simply be the reappearance of God in masked form (*M*, 48).

It is a curious science indeed which needs God, not only for it to be invented (several scientists have thought Him necessary for this purpose), but in order to function. For where else would language get its force, if it does not come from either meaning or the world of things, and still less from feelings, drives, or desires?

To sum up, what would it take for Lacan's proposed little machine to function?

1. The linguistic formalization of the phoneme would have to lead to a true algebra, which is not the case. Or again, linguistic binarism, which makes it possible to draw up sets of tables marking presence and absence, and which has to do with taxonomy, would have to be identified with the cybernetic binarism that makes it possible for a machine to operate; but once again, this is not the case. Levi-Strauss, who would have liked to submit the future of the social sciences to "the mathematical methods of prediction which have made possible the construction of electronic calculating machines,"[8] a few pages later acknowledges that phonology leads not to the construction of a machine, but to the construction of a table similar "to the table of elements which modern chemistry owes to Mendeleev."
2. This linguistic formalization, supposedly of a mathematical order, yet which holds only for phonemes, would have to be valid at the lexical level, and would then also have to be applicable to messages, that is, to sentences, whether implicit or explicit. But sentences cannot be pure signifiers, since the meaning of a sentence is not independent of its construction.
3. Supposing that the first two conditions were met, a machine, in order to function, has to be endowed with energy; it is therefore necessary to invoke a form of transcendence, or a God. All these things, then, would be necessary for psychoanalysis to become a science!

But let us nevertheless assume that the machine actually works. We then enter a world of necessity extolled by Lacan in a famous passage from his "Discourse":

> Symbols in fact envelop the life of man in a network so total that they join together, before he comes into the world, those who are

> going to engender him "by flesh and blood"; so total that they bring to his birth, along with the gifts of the stars, if not with the gifts of the fairy spirits, the design of his destiny; so total that they give the words which will make him faithful or a renegade, the law of the acts which will follow him right to the very place where he *is* not yet and beyond his death itself; and so total that through them his end finds its meaning in the last judgment where the word [verbe] absolves his being or condemns it—except he attain the subject realisation of being-for-death (*E*, 279).

Faced with this implacable world, what is there left? After having backed [misé sur] desire, so that living beings were not annihilated in symbols, and after having briefly evoked "a psychoanalysis [in which] what is at stake is the advent in the subject of that little reality which this desire sustains in him with respect to symbolic conflicts and imaginary fixations" (*E*, 279), Lacan then comes to expound at length on the actual impasses of the subject's speech in his relation to language. These impasses concern madness, symptoms, and the objectifications of discourse in which man forgets his subjectivity, his own existence, and his death (*E*, 279–83). Although at one point an incontestable creative subjectivity is mentioned, it is quickly turned into its opposite, as often happens in revolutions (*E*, 283). Such catastrophic analyses are to be attributed not to the pessimism of their author, but to the most elementary logic: If the Symbolic is cut off from its social foundations, if it is forced to become autonomous in order to meet the needs of the scientific cause, no society at all could possibly be salvaged from it. In other words, since language is already everything, one has only to empty life, desire, and later anxiety and sexuality (which will have to be included in the field) of their content. Since we are dealing with Freudian psychoanalysis, it will certainly be necessary to try to slip all of these elements back into the system, but Lacan will regularly re-expel them, so as to find himself once again face to face with nothing.

Lacan will settle for nothing less than inserting psychoanalysis into the movement of science. He does not bluntly say that psychoanalysis is a science, nor even that it is going to become one; he leaves the conclusions to the reader, who will be given to believe that the way to formalization is open. Starting with a vague definition that goes back to Plato, Lacan moves on to the conjectural sciences to promise "a precisely defined approach to our own field," thanks to the mathematicization introduced by phonematics. He then announces a general theory of the symbol "in which the sciences of man will again take up their central position as sciences of subjectivity" (*E*, 285), giving him a further opportunity to evoke the model of all possible sciences (mathematical physics), and, aided by a trio of double negations, to claim a virtual closing of the gap between physics and psychoanalysis:

> Here the opposition which is traced between the exact sciences and those for which there is no reason to decline the appellation 'conjectural' seems no longer an admissible one—for lack of any grounds for this opposition [Ici *n*'apparaît *plus* recevable l'opposition qu'on tracerait des sciences exactes à celles pour lesquelles il *n*'y a *pas* lieu de *dé*cliner l'appellation de conjecturales: faute de fondement pour cette opposition] (*E*, 286).[9]

It cannot be said that Lacan is not extremely prudent in his formulations, that he does not consistently and calculatedly shun everything that could not fail to be raised as an objection against him, so much does he deny and strive to conceal an incoercible desire to have us believe he has achieved something which he clearly knows he has not achieved, so that he himself can believe a little more firmly in the efficacy of his subterfuges, or in the force of his proofs, which supposedly emerge from the mere juxtaposition of disparate elements. For, after having posited the impossible nonproximity of the exact and conjectural sciences, in which a certain modesty—merely the obverse of a passion—lies coiled in wait, he nonetheless resorts to the double strategy already discernible in the 1936 text: He erodes the pretensions of physics ("its relationship to nature remains no less problematic," "it is merely a mental fabrication" defined more by measurement than by quantity, *E*, 286), and plays up the attainments of psychoanalysis in the formalization of subjective time. For example, "to have tried to demonstrate in the logic of a sophism the temporal sources" of human action, is supposedly akin to "the mathematical formalization which inspired Boolean logic;" and the structural logic which applies to phonemes is supposedly transposable into the structurations of language "in the interpretation of resistances and the transference" (*E*, 288).

If Lacan discovered *the* Symbolic through reading Levi-Strauss, he was able to assimilate the triad of symbolic system, linguistics, and formalization,[10] with the whole apparatus then aligned with the unconscious, because his own previous research had made him receptive to the idea. In his "Au-delà du 'Principe de réalité' " ["Beyond the 'Reality Principle' "], as in his "The Mirror Stage," he essentially sought to give precise outlines to Freud's "other scene," wanting thereby to extricate psychoanalysis from the vagueness of psychological description in order to provide it with a formal or structural basis. His use of the imago, assimilated to the Gestalt, was supposed to rescue psychology from sheer behavioral or subjective diversity. By aligning Wallon, Freud and the gestaltists, he had created a background, a stable basis whereby, through phenomenological reduction, analysis could account for the indefinite multiplicity of the visible and the sensible. In spite of his exceptional critical gifts, evident for example in his lecture of the same year on "Le mythe individuel du névrose," his ambition to turn psychoanalysis into a science could not be realized, since it was impossible for him to make his way out of the register of description or deduction that way, however vigorous or sophisticated the latter might be. It is understandable, then, that Levi-Strauss's impressive feat of uniting ethnology, linguistics, mathematics, and psychoanalysis in the same discourse should find an echo in him and arouse phantasms of science. For this combination of disciplines was a formidable war machine committed to an invasion of the culture (something that Lacan immediately sensed), but first and foremost it was a veritable magical solution after years of research and groping about. In his "Discourse of Rome," there is a lot more than just a "trace of enthusiasm" (*E*, 229), a "trace" that he says is regrettable, and without any mitigating circumstances. The "Discourse" is in fact a veritable trumpet blast designed to bring down the banners. In the introductory text published in 1966, he is still quite insistent, and there is no sign of even the slightest retraction: "We are trying for an algebra which would answer, in the place thus defined, to what, for its part, the kind of logic called symbolic accomplishes when it delimits the laws of mathematical practice" (*E*, 233).

The enterprise would thus seem to be going rather well. But supposing that it is, or even that it has been fully successful, the question is, at what price? To answer this question, it is initially necessary to refer to Levi-Strauss's critique of Mauss's interpretation of the notion of the "mana." For Lacan's implicit acceptance of this critique will decisively affect the way in which he subsequently conceives of psychoanalysis, and will make it seem necessary, in the name of science, to construct an enclosure for it.

Levi-Strauss reproaches Mauss with having sought to translate the notion of mana in terms of feelings, wishes, and beliefs in order to make the Polynesians' encounters with secret powers and mysterious forces comprehensible to us. Levi-Strauss considers all realities of this order to be "from the standpoint of sociological explanation, either epiphenomena or mysteries, in any case objects extrinsic to the field of investigation" (p. XIV). The social, which Durkheim and Mauss identify with the sacred and which was the object proper of sociology, must be eliminated and replaced by language—that is, by the symbolic system.

> Levi-Strauss thus proposes a rationalist reform of the Academy's doctrine. For him, the sacred no longer refers to an experience at all—an experience in which man puts to the test a higher power which threatens to destroy him—*but to an effect of language,* to the presence in discourse of symbols x, y, or z which betray the imbalance between what we *can say* about the universe because we know certain things and what we *could say* about the universe if we knew more. These x's and y's are the marks of our present inadequacies, as well as signs of our future discoveries. In the final analysis, the sacred is reduced to being simply the indication of the distance separating us from omniscience.[11]

This procedure allows Levi-Strauss to construct a science, but subject to the condition of eliminating its very object, simply because the object presents itself in a mysterious form, confusing and difficult to decipher. Whereas "Mauss's conception of science does not prevent him in advance from envisaging his object (. . .), in Levi-Strauss, by contrast, there is a law which precludes science from encountering anything in its field of investigation apart from what it is required to encounter by virtue of its principles and methods, namely, the calculable."[12] The victim of these principles is precisely the mana which was meant to be explained, the mana that Mauss aligned with the *physis* and *dynamis* of the Greek alchemists, words which we have to translate as "force" or "power." The new ethnological science and, before too long, psychoanalysis, have to ignore force and power, which also means ignoring everything that pertains to feelings, wishes, and beliefs—not to mention anxiety and the imagination.

Lacan certainly sensed the difficulty, since he refers to the *Introduction* in the following terms:

> Identified with the sacred *hau* or the omnipresent *mana,* the inviolable Debt is the guarantee that the voyage on which wives and goods are embarked will bring back to their point of departure in an unfailing cycle other women and other goods, all bearing an identical entity: what Levi-Strauss calls the "zero-symbol," thus reducing the power of Speech to the form of an algebraic sign (*E,* 279).

But the problem lies precisely there: How is it possible to attribute power to an algebraic sign? We are brought straight back to the equivocal principle conflating algebraic symbolism and social symbolism.

In spite of having sensed the difficulty and having claimed to resolve it, through his passion to found a science Lacan is nonetheless led to refusing to consider the object and the means of the analytic experience. Here he gets out of trouble through irony: "The lived reaction of which they show themselves to be so fond (. . .). Keep going in that direction and no doubt a mutual sniffing will be the last word in the transference reaction" (*E,* 267); and in his lecture of the same year, programmatically entitled "The Symbolic, the Imaginary and the Real," he does it through positing an inviolable limit. He admits that

> in analysis, there is a whole section of the real in our subjects which eludes us, something we deal with all the time (. . .) this element of immediacy, of weighing someone up, of assessing his personality, of deciding whether or not the subject has some substance. This is something which constitutes the limits of our experience. This is the sense in which one can say, in order to pose the question of what is at stake in analysis, "What is it?" Is it this real relationship with the subject which needs to be recognized?—I mean in a certain way and according to the means of recognition at our disposal. Is this what

> we're dealing with in analysis? Definitely not. Incontestably, it is something else. Indeed, there we have the question that we and all those who try to theorize the analytic experience ceaselessly ask ourselves, (. . .) the question of the irrational character of analysis. (. . .) From there to thinking that analysis itself operates in a certain register, in magical thought, naturally, is but a single step, a step quickly taken when one does not begin with and decide to stick to, from the outset, the primordial question: What is this experience of speech?—and to simultaneously pose the question of the analytic experience, the question of the essence of speech, and of its exchange.[13]

This long oscillation is necessary because the question cannot be sidestepped; but it is mentioned here only in order to better prepare for its expulsion, when it vanishes and is replaced by the role language plays in psychoanalysis. In this text, which should be cited in its entirety because it is a model of its genre, Lacan gives the impression of taking the web of experience into account, but merely in order to exclude it, under the pretext of it being an inviolable limit. It has to be said, however, that if there is an uncrossable line here, it is simply because it has been decided in advance that this aspect of experience lies outside the field, and once again all in the name of science. This is certainly a strange form of reasoning; after acknowledging that this "element of immediacy" is ceaselessly encountered in the analytic experience, Lacan then jettisons it because it is unknowable, by virtue of the very means and method employed, and by virtue of the confusion posited at the outset between the statement that speech is the means of analysis, and the claim that speech alone comes into play in analysis. This confusion traverses Lacan's entire *oeuvre,* but was only necessary because he so badly wanted to found a psychoanalytic science.

In the years that follow, Lacan keeps the same model of science in mind. In his 1956 seminar on *The Psychoses,* for example, he writes:

> We are situated in a field distinct from that of the natural sciences, and as you know, to call it the human sciences is not the whole story. How to demarcate them? To what extent must we move towards the ideals of the natural sciences?—I mean such as they are, laid out for us. Physics, for example, which is what we're dealing with here. To what extent are we unable to tell them apart? Well, it

> is in relation to these definitions of signifier and of structure that the appropriate border can be marked out (*P,* 208).

But before seeing how he responds to these issues, we need to ask ourselves how they came to be raised at this point in the seminar.

A sudden interjection from the audience impels Lacan to interrupt his reading of Schreber:

> Fortunately, you're not the only one in the Society of Psychoanalysis. There is also a woman of genius, Françoise Dolto, who has shown us the essential function of the body-image, and has enlightened us on the way in which the subject relies on it in his relations with the world. We are very glad to find a substantial relation here, onto which the relation to language is no doubt sewn (sic), but which is infinitely more concrete (*P,* 183–84).

The show he puts on in response to this objection is that of a great rhetorician or a true politician. He starts by disarming his interlocutor, suggesting that there has been some misunderstanding, and that this is perfectly normal, since "I pursue this discourse with an express, if not absolutely premeditated intention of offering you the opportunity of not quite understanding it." Besides, he says, if I made myself readily understandable, "the misunderstanding would be irremediable," because you would then be imprisoned in certainty (*P,* 184), and for that reason I perpetuate misunderstanding.

In fact, it is not at all a question of misunderstanding, and what follows this exchange shows—but only after a long detour—that the objection raised by certain members of the audience retains all of its force and has even more *raison d'être,* and its exclusion in the name of science strongly resembles and repeats in its own way the same move in Levi-Strauss.

The argumentative procedure is clear: You start by conceding everything to your opponent in order to put him to sleep, all the while planning to reject everything he says: "That this pre- or even extraverbal communication is permanently there in analysis is not in doubt, but it is a question of determining what constitutes the properly psychoanalytic field" (*P,* 184–85). In other words, nonverbal communication is permanently there in analysis, but I'm going to show you how not to have to bother with it. The trouble is, no sooner is it expelled from one place than it reappears in another. The repressed returns, even in Lacan.

Lacan claims, of course, that "analysis has shed an immense amount of light on the preverbal," it "has led us to explore the imaginary world" (*P*, 185), which is simultaneously not all that unfamiliar and yet remains unfathomable (*P*, 186). In order to get rid of this domain, in which "the essential findings" of psychoanalysis are not located, Lacan pillories it with the qualifier "preconscious." Now, as we know, analysis is interested first of all in the unconscious, as defined by the Lacanian interpretation of Freud: "Every analytic phenomenon, every phenomenon participating in the analytic field, in the analytic discovery, in what we deal with in symptoms and neurosis, is structured like a language" (*P*, 187).

This is not the place to discuss his use of the term "preconscious," nor his conception of the unconscious (a conception which we now know descends directly not from Freud, but from Levi-Strauss). The main thing is to highlight Lacan's strategy for dispensing with the interjection: The preverbal, the Imaginary or the imagination are encountered in analysis, but they do not form a part of the field proper of analysis, because, according to the well-known question-begging move, this field is defined by the signifier, "the sign of an absence," "a sign which refers to another sign, a sign which is structured as such in order to signify the absence of another sign, in other words, in order to be opposed to it in a dyad" (*P*, 188). He thereby prefigures what will later be called structural analysis: "This characteristic of the signifier marks in an essential manner everything which belongs to the unconscious order" (*P*, 188).

Then comes an argument endeavoring to show that, if symptomatic manifestations are colored by the imaginary, they can be understood only through their symbolic identifications—that is, from the standpoint of the Other. Similarly, human sexuality can be grasped in its specificity only when situated in this same relationship—whence the spiteful assertion to which one can no longer object: "If the recognition of the sexual position of the subject is not linked to the symbolic apparatus, then analysis, Freudianism, have nothing further to do but disappear; they mean absolutely nothing" (*P*, 191). Obviously, with it being so hard to establish, no mention is made of this link to the Symbolic.

Lacan is, nonetheless, going to run into a fundamental difficulty. After having asserted that

> The Symbolic provides a form in which the subject is inserted at the level of his being. It is with the advent of the signifier that the subject recognizes himself as being this or that. The chain of signifiers has a fundamental explanatory value, and this is what the notion of causality is all about (*P*, 201–2),

he is forced to admit:

> But the fact of their [the subjects] individuation, the fact that one being emerges from another, [un être sort d'un être], is explained by nothing in the Symbolic. The symbolism is all there to affirm that creatures do not engender creatures, that creatures are unthinkable without a fundamental act of creation. And nothing in the Symbolic explains creation (*P*, 202).

And then a little further on he says:

> There is something which is radically unassimilable to the signifier. It is, quite simply, the singular existence of the subject. Why is he there? Where does he come from? What is he doing there? Why is he destined to disappear? The signifier is unable to give him the answer, for the very good reason that it situates him, exactly, beyond death. The signifier considers him already dead, and immortalizes him in his essence.

What, then, is this "fundamental explanatory value," this "very notion of causality," which is incapable of explaining individuation, the singular existence of the subject—in a word, his life—and which, into the bargain, has to turn him into a dead man in order to understand him? In individuation, singular existence, and life, is it not quite simply subjectivity,[14] which is overlooked?—subjectivity, with which analysis rightly concerns itself (even if it does not seek to answer the "why" of existence, which is something else altogether) when it takes an interest in the preverbal, the Imaginary, and the imagination, which do not pertain solely to the preconscious, but are, for Freud, characteristics of the dynamic unconscious.

Lacan is in any case forced to abandon this domain to what is inessential in analysis, because he wants to found psychoanalysis as a science—a science, to be sure, not identical with but nevertheless comparable to the exact sciences. Now, it is precisely at this stage in Lacan's discourse that the above questions spring up again. He introduces them with a definition of structure—"a group of elements forming a covariant set" (*P*, 207)—and an alignment of the signifier with structural analysis, an alignment too vague to object to, but insistent enough to have us believe that the essential link has been established:

> To be interested in structure means being unable to neglect the signifier. In structural analysis, as in the analysis of the relation between the signifier and the signified, we find group relations based on open or closed sets, but essentially involving reciprocal reference. In the analysis of the relation of the signifier and the signified, we have learned to place the emphasis on synchrony and diachrony, and this is also found in structural analysis. In the final analysis, when examined closely, the notions of structure and the signifier appear inseparable. In fact, whenever we analyse a structure, it is always, at least ideally, a question of the signifier. What satisfies us most in a structural analysis is as radical a detachment of the signifier as possible (*P*, 208).

And then comes the series of questions:

> We are situated in a field distinct from that of the natural sciences, and as you know, to call it the human sciences is not the whole story. How to demarcate them? To what extent must we move towards the ideals of the natural sciences?—I mean such as they are, laid out for us. Physics, for example, which is what we're dealing with here. To what extent are we unable to tell them apart? Well, it is in relation to these definitions of signifier and of structure that the appropriate border can be marked out (*P*, 208).

How does he respond to these questions? Certainly not directly, for it would then become obvious that physics and psychoanalysis have nothing in common. In order to give the impression that the two disciplines do overlap in spite of their differences, he will have to lead the reader astray in detours, and redouble the confusion.

He starts with physics: "In nature, no one uses the signifier to signify. This is what distinguishes our own physics from a mystical physics, and even from ancient physics. (. . .) The signifier is there, however, in nature, and if we weren't after the signifier, we would find nothing there at all. To isolate a law of nature is to isolate a nonsignifying formula" (*P*, 208). How should these statements be construed? If the accent is placed on "no one uses the signifier," you end up with the age-old idea that there are no voices or spirits in inanimate objects. Only the scientist will give voice to the signifier that is, nonetheless, in nature. If the emphasis is placed on "in order to signify," you are already suggesting that the signifier signifies nothing and paving the way for "to elicit a nonsignifying formula," which recalls the theory of the autonomy of the signifier, highlighted by the sentences that follow: "You would be wrong in believing that Einstein's little formulae, which align inertial mass with a constant and some exponents, have the slightest meaning. This is a pure signifier." We thus seem forced to conclude as follows: Physics is a science because it does not claim, as mystical physics does, to uncover meaning in nature, and because it elicits pure, meaningless signifiers. The way is then open to some fruitful alignments with psychoanalysis.

But for such a conclusion to be tenable, a few obscurities would first need to be cleared up. First, how is it possible to claim that there are signifiers to be discovered in nature if no one uses signifiers in order to signify? In other words, how can a signifier exist if no one speaks or has spoken? Second, how can the signifier be at the same time a word that is interchangeable with metaphor or metonymy, both of which can be attributed a certain autonomy with respect to meaning, *and* a sentence stating something.[15] Now, "Einstein's little formulae," even if they take an elliptical, alphabetical form, are true sentences. What the equation says can be clearly stated, in perfectly good French. In other words, these formulae do have a meaning. Lacan has therefore confused Saussurian arbitrariness (signifiers are defined by their differential relationship in the language), and algebraic arbitrariness (where the letters used in a formula serve to figure magnitudes whose relationships are defined by an equation. This mutual definition has nothing to do with the morphological arbitrariness of words or Saussurian signifiers; it is a question of the physicist's redefinition of one concept in function of others).

If we do not accept the willful confusion at work in these pages, we are forced to conclude that physicists do not employ the Saussurian or Lacanian signifier, and that the formulae they establish *do* have a meaning. There is no question, therefore, of using this method to mark out any kind of border between the two disciplines.

Let us nevertheless assume that it has been demonstrated that the bar between signifier and signified is also found in physics; how will the psychoanalytic field be defined in opposition, since "the appropriate border" (*P*, 208) has still to be marked out? Essentially it is defined through the Symbolic, reduced to the signifier that signifies nothing: "Our starting point, the point to which we always return, since we will always be at the starting point, is that every true signifier is, as such, a signifier which doesn't signify anything" (*P*, 210). This doesn't prevent him from stating the obvious a little further on: "this signifier is always brought into play in signification" (*P*, 223), and "of course, by definition, the emergence of a pure signifier cannot even be imagined" (*P*, 225). A choice surely has to be made between these conflicting claims, but, since the entire theoretical edifice would collapse without the autonomy of the signifier, in accordance with good Lacanian logic, we have to take it as a given.

We therefore find ourselves confronted with two formulae, one defining the field of science, where "no-one uses the signifier to signify," and the other characterizing the field of psychoanalysis: "Every true signifier is, as such, a signifier signifying nothing." Obviously, these two formulae are never placed in opposition, and still less are they explicitly presented as equivalent. Yet a choice has to be made between them. They either mean that, in physics and in psychoanalysis, the signifier is cut off from the signified or from signification, which blurs the border between the two disciplines; or they mean that, in physics, the signifier is simply not used, whereas it constitutes the very basis of psychoanalysis—in which case physics and psychoanalysis have strictly nothing to do with each other. Inevitable as it is, this is just the kind of conclusion that Lacan carefully avoids, because it would render his constructions as useless as they are unacceptable.

As he lets the equivocations accumulate, it is enough as far as he is concerned that the separation of the signifier and signified appear to his audience to be subjacent to both fields, in such a way that they come to believe in their proximity. But such a belief will be entirely the audience's affair, since Lacan will be able to deny having suggested anything of the kind.

Lacan can now return to the major difficulty, which he had stumbled over a moment earlier: The Symbolic does not explain individuation, the singular existence of the subject, or creation. To resolve the problem, he introduces the term "subjective" into the newly-formed context (a term that he had deliberately kept in reserve), and couples it with the word "Real."

It is impossible to outline in detail here the multiple implications of the term "Real" in this seminar, much less its implications in Lacan's *oeuvre*. That will come later. Suffice it to point out that the Real is intrinsically linked to the epistemology of science: The Real is the reality uncovered and formulated by the laws of physics, the reality that lies beyond perceptible reality; it is thus essentially of a mathematical order. By aligning these two notions of "subjective" and "real," Lacan's strategic aim is twofold; on the one hand, he aims to account for what science excludes and psychoanalysis includes and, on the other, to found psychoanalysis as a science.

The mirage that considers the subjective and the objective to be opposed has to be denounced straightaway (*P*, 210). The subjective "is not in the domain of the speaker. It is something we encounter in the Real" (*P*, 211). The scientific Real, however, has to do with objectivity, a difficulty that will be surreptitiously gotten rid of: "No doubt the Real in question is not to be understood, as it habitually is, as implying objectivity—a confusion incessantly found in psychoanalytic writings" (*P*, 211). After the words "implying objectivity," one would expect the sentence to end with a formulation like: "as is the case in physics," since this is what science involves: As much as possible, it seeks to eliminate the observer and thus the subjective from its findings. But this is precisely the kind of conclusion that would completely destroy the ensuing argument. It therefore means saddling the poor psychoanalyst—and in the name of a supposed confusion on his part—with one of science's constant claims; that way Lacan appears to be opposed not to science, but to bad psychoanalysis.

The subjective is now to be understood as follows: "The subjective appears in the Real to the extent that we have before us a subject able to use the signifier, the play of the signifier. And able to use it as we do—not to signify something, but precisely in order to outwit what can be signified" (*P*, 211).

These two sentences say two decidedly different things. On the one hand, the play of the signifier refers to something that belongs to the mathematical real. This is why it is defined soon after as "a correlative system of elements, each placed synchronically and diachronically in relation to the others" (*P*, 213), which is an allusion to structure or structural analysis. On the other hand, the distinction between the signifier and the signified is given a completely distinctive twist, since it is referred to the analytic experience, where speech always proves to be marked by the stumblings of double meanings, which can seem to be, or be said to be, deception. We are thus a long way from science.

But this is not the case if we accept the—to say the least—slippery reasoning that follows Lacan's evocation of the deceptive signifier: "This is so essential that it is properly speaking the first step in modern physics." One is left goggle-eyed on reading the explanation for this: "The Cartesian discussion of a deceitful God [Dieu trompeur] is the one step it is impossible to avoid in any effort to found a physics, in the sense in which we understand this term" (*P*, 71). Thus Descartes' discussion of "God as a God who cannot deceive us" (*P*, 211) now becomes "the Cartesian discussion of a deceitful God," the basis of physics. There would be a great deal to say about this conception of the founding of physics (since physics is certainly not founded on this), but what is striking here is that the God incapable of deceiving us suddenly becomes the God who can. And yet there is nothing at all surprising about this. Physics and psychoanalysis have to be brought closer together: The former was founded on a God who cannot deceive us, and a God who can is needed to found the latter. If it were stated clearly and explicitly, the opposition between physics and psychoanalysis would be irremediable; Lacan therefore effaces the problem by talking in terms of "the Cartesian discussion of a deceitful

God." This is, after all, perfectly acceptable, since the word "discussion" leaves open the possibility of a negative conclusion. Once again, Lacan is suggesting that something that has nothing to do with the problem in question (deception in psychoanalysis—which, moreover, is not deception, but equivocation—and a God incapable of deception) has everything to do with it.

He can then go on to formulate his response to all the issues raised earlier:

> The subjective is for us what distinguishes the field of science in which psychoanalysis is based from the overall field of physics. It is the instance of subjectivity as present in the real which is the essential source of what is new in what we are saying here, for example, when we distinguish the seemingly natural series of phenomena that we call the neuroses and the psychoses (*P*, 211).

We are thus led to believe that the sought-after border between the two disciplines has been definitively drawn. But the clouds gather once again when we read: "Are the psychoses a series of natural phenomena? Do they enter the field of natural explanation? I call *natural* the field of science in which no one uses the signifier in order to signify." Physics thus returns to the scene, only this time without us being entitled to any further explanation. All this shuttling back and forth is necessary because Lacan wants to distinguish between physics and psychoanalysis by reintroducing subjectivity, but at the same time he wants to give psychoanalysis a force akin to that of physics. To do so, he has to have us believe that the subjectivity peculiar to psychoanalysis has been accounted for, while draining it of its content so as to make it manipulable, in the same way that a scientific object is manipulable. In the wake of all these contradictory moves, he feels he can claim to have presented "a scientific definition of subjectivity (. . .) on the basis of the possibility of deflecting the signifier towards signifying, and not significative ends, that is, a signifier expressing no direct relation with the order of the appetites" (*P*, 214).

One can only admire in this passage the manifold skill with which Lacan succeeds in turning round a desperate situation, and the way he successfully manipulates superficially similar notions, which finally prove nothing, but which the audience is nonetheless invited to view as conclusive. Equally admirable is the way he always manages to avoid using formulations that do not carry or precede their own negation; the way he succeeds in creating a whole universe of notions having all the trappings of precision, and which, when combined with others, have some chance of being mistaken for rigor; and finally, the way he avoids meeting difficulties head on by diluting them, bogging them down in detours where they are lost and forgotten. Aided by this impressive array of subterfuges, he has supposedly founded a science on the basis of what science excludes.

This does not mean that his project has finally been realized, since the question of the scientificity of psychoanalysis will go on running into the question of the relationship between psychoanalysis and science. The scientificity of psychoanalysis is dependent on its accomplishments in the field of science, and Lacan advances with a great deal of prudence in this domain. He constantly avoids formulations which would allow his thought to be reduced and its end result clearly grasped. In a sense, he proceeds with incontestable rigor, but a rigor which, paradoxically, only holds together thanks to his systematic use of equivocation, covering up the leaps in his reasoning as well as all the hasty assimilations. His style is wholly oriented towards the juxtaposition of terms which have to be seen as related for a conclusion to be reached, but which the audience itself is precisely forbidden to relate for fear of its discovering the incoherence, the futility of the argumentation, indeed the exemplary trickery. In this relentless form of self-contradictory rigor, it is impossible not to recognize the mark of the most inventive genius, and the surest sign of its bewitching power.

In his 1965 text entitled "Science and Truth," Lacan gives a further demonstration of the art of allusion and suggestion, which he finds indispensable for saying and not saying something at the same time. Analyzing this text will once again allow us to grasp how he proposes to found psychoanalysis as a science.

The text opens on this very note: "Can we claim that last year we founded the status of the subject in psychoanalysis?" (*E*, 855). It should be noted that this is a question. Indeed, Lacan carefully refrains from any assertion that would result in him encountering a mind exacting enough to be likely to ask for explanations, a strategy that allows him to answer the question in the most modest and incontestable fashion: "We have succeeded in establishing a structure which accounts for the phenomenon of splitting, *Spaltung*, which the psychoanalyst encounters in his practice" (*E*, 855). It is true that this division of the speaker is a constantly encountered fact of analytic practice: The analysand lets himself go, saying something other than what he wanted to, often not recognizing himself in what he says. But the words "status" and "found" contain an ambition altogether different from providing a justification of analytic practice; they are aimed at "the birth of a science," over which "a reduction which properly constitutes its object" (*E*, 855) will have to preside.

Thus the status of the subject must not only be such that it accounts for the empirical, but a reduction (as the epistemologists say) has to be performed on it, so as to constitute the object of the emerging science. The words "subject" and "object," which in their adjectival forms had to be kept at a distance from each other, are here disturbingly juxtaposed. It could be said that the entire strategy of this text is already contained in this one move. What must be kept out of sight, or at least not be explicitly stated, and which is nonetheless the nerve center of the reasoning, is the extreme proximity, indeed the identity, of subject and object. Yet from the very first page, through all the sliding from the psychoanalytic subject to the scientific object, we are alerted to the fact that in this move we have what is essential to what has to be established and at the same time concealed. For psychoanalysis to become a science, the subject has to be turned into an object; but since this would mean losing his status as a subject, the operation has to be covered up.

It is then that, like a rising star, the grand model appears: "For I am not sure that [epistemology] has thereby [through reduction] fully accounted for that decisive mutation which, by way of physics, founded *science* in the modern sense, a sense which is posited as absolute" (*E*, 855). Reduction is thus not enough to account for the radical change brought about by science; a "modification of our subject position" (*E*, 856) was also needed. Consequently, Lacan's effort to found the new science will consist of carrying out a reduction on the position of the subject. But he doesn't make his move straightaway, since it would be too visible and thus too vulnerable. When establishing the relation between psychoanalysis and the subject of science, it is better to take an indirect, unassailable route.

We are successively referred to the Cartesian *cogito,* which is supposedly the source of modern science, because Lacan finds in it a version of the "experimental division of the subject" (*E,* 856) in analysis; to Freud, who prefigures "what structuralism has since allowed us to elaborate logically, namely the subject, and the subject grasped in a constitutive division" (*E,* 856), and to Freud again, whose discovery would have been unthinkable prior to the century of science and who never broke away from "the ideals of scientism" (*E,* 857).

All this is meant to have us think that there are intrinsic links between psychoanalysis and science. But what are these links? Lacan seems to take a step forward and accept the elementary question of the incompatibility of psychoanalysis and science, since the latter leaves the subject to one side: "To say that the subject we deal with in psychoanalysis is none other than the subject of science may seem paradoxical" (*E,* 856). Instead of clearing up this paradox, though (and explaining what, in this passage, "the subject of science" might actually mean), it is better to launch into some diversionary tacts—by noting, for example, a "dishonesty elsewhere called objective: but this is for lack of nerve and comes from failing to locate the malfunctioning object [l'objet qui foire]" (*E,* 859)—a laughable remark, since the whole idea here is precisely to make the true object of psychoanalysis malfunction. He then mounts an attack on the position of the psychoanalyst, which "precludes the tenderness of the beautiful soul [belle âme]," for the psychoanalytic field has to be emptied of all relations to the affective. Then comes his response to the above paradox: "If it remains a paradox to say it, it is also perhaps still one and the same [subject]" (*E,* 859), meaning that if the subject we work with in psychoanalysis is the subject of science, this fact assumes a bracketing of everything that would ordinarily be called subjectivity. This kind of argumentation is then emphasized in the paragraphs that follow: Seeking to embody the subject in childhood or in the primordial is a theoretical error. Thus, psychoanalysis deals only with a disembodied subject: "Only one subject as such is admitted in psychoanalysis, namely the one able to render it scientific" (*E*, 859). Are we dealing with question-begging here, or with a vicious circle? Psychoanalysis will be scientific if it deals with a scientized subject. Fine—but the question that is never dealt with is that of establishing whether this subject will still be the same as the one encountered in psychoanalytic practice.

The counter-attack continues, this time with a rejection of the human sciences: "There are no human sciences, because the man of science does not exist–only its subject exists" (*E*, 859). We are forced to conclude that psychoanalysis as a science addresses itself to a dehumanized subject. This does not prevent him from criticizing psychology, which, having become a science, is merely the slave of technology, or from predicting its failure when it seeks to isolate the "creator in science" (*E*, 859), whereas some pages later it will be said that this creator and his dramas elude science (*E*, 870).

After this trial-like ground-clearing [déblayage en forme de procès], the sciences that rule out the "illusion of the archaic" or "psychologisation" (*E*, 860)—in other words, the sciences that open the way to scientific psychoanalysis—can emerge. In fact, one scarcely sees what hopes they could possibly raise, since they remain in the same state now as they had been 12 years earlier (*E*, 284–88). They have not given rise to a single development, they are still just as vaguely described, yet they are nonetheless supposed to have the virtue of convincing us that, with their help, a science of the subject will be possible. An example of such a science is "game theory (. . .), which has the advantage of the entirely calculable nature of a subject strictly reduced to a matrix of signifying combinations" (*E*, 860). What better way of saying that we are in fact dealing with a disembodied and dehumanized subject without individuality, a subject who, by all appearances, could be reduced to a pure object—in this case, to numbers? But this is exactly what Lacan has to avoid saying.

He then invokes linguistics, whose marvelous gift it is to go "a long way in the elaboration of the effects of language, since a poetics can be constructed here which owes nothing either to references to the mind of the poet or his embodiment" (*E*, 860). This is a form of linguistics that considers poetry as a text without reference to the subject who

the overhasty extrapolations, since these were never anything but excrescences whose validity could be doubted without this uncertainty having repercussions on the context as a whole.

This was no longer the case once Lacan set out to follow in Levi-Strauss's footsteps, and to put psychoanalysis on the same footing as ethnology, linguistics, or logical formalism. He then foisted on himself a collar ill-suited to his discipline.

For the Lacanian machine to function would have required that the principal reasoning behind it be something more than mere sophistry. This reasoning can be summed up as follows: Since the psychoanalytic method draws on language alone, and this method provides access to the unconscious, the unconscious is structured like a language, it *is* a language (*P*, 20), it *is language* (*E*, 866). This is sophistry because the instrument of research is confused with the object of the research.[16] Just because certain celestial bodies can be studied only by means of a telescope does not mean that celestial bodies have the same nature as a telescope. And it is sophistry because two decidedly different assertions should not be confused: That language is an instrument of psychoanalysis, and that language is the *only* instrument of psychoanalysis. This second claim, ceaselessly repeated by Lacan, is quite simply false, since the transference plays a large part in psychoanalysis, and, in spite of the efforts in this direction, it is irreducible to either language or knowledge [savoir]. And finally, it is sophistical because the term "language" would need to be defined. Is it a code? Is it a message? A code is structured, it is a language *in potentia;* but it cannot be identified with the message, which is a text, even if reduced to a single word. And a text is oriented, it has a meaning, a signification; in no way is it open to being assimilated to a combinatory, and no truly algebraic formula has ever been found to emerge from it.

Lacan's machine is thus inert, unable to function; it may be a work of art, to be contemplated, but it is unusable. If you do use it, or think that you do, it is only because something else is driving it. For it is clear that, if you really want to believe that the apparatus works, you are forced to acknowledge that it expels the very content it was supposed to transform. Lacan clearly recognizes this: Life, individuality, imagination, and affect are unassimilable by the signifier as he defines it. As for subjectivity, we have seen that it was of interest to the new psychoanalysis on condition of its being emptied of all flesh and blood, all humanity, so as to be reduced to the dimensions of an "x" on a graph.

What remains of the Freudian unconscious? We saw that the term was avoided until a way had been found to reduce it to a pure combinatory of signifiers, in spite of the claim that we were still dealing with what Freud discovered. Supposing that this combinatory exists, and that it accounts for what speaking beings actually say and the way they behave, is there any justification for identifying it with the unconscious, and for using the same term? For if we have thereby been introduced to science, we can ask ourselves, for instance, whether it would occur to a scientist to claim that the algebraic formulae he had produced in order to explain a certain phenomenon had previously been unconscious, and for many people still are. Such a way of expressing yourself would strictly speaking have no meaning, for the unknown is not to be confused with the unconscious.

Otherwise psychoanalysis would have to be reduced to an enterprise that simply produces a form of knowledge [connaissance], a kind of ethnology applied to individuals. In fact, it was in this direction that Lacan was headed, since, for him, "the only thing worth the trouble" is "what it means *to know*" (*I*, 120).

But, since Lacan wanted to found psychoanalysis as a science, it was enormously important for him not to deprive himself of Freud's authority, which is why he ceaselessly claimed to be merely following him and interpreting him as he was supposed to be interpreted. To do so, he had to reintroduce all the Freudian metapsychological concepts into his theoretical elaboration—a Sisyphean task, which he nonetheless did not balk at. But how can it be demonstrated that the transference, desire, the drive, anxiety, and affects do indeed have a place in the new science? In other words, how can it be proven that the system of the signifier and the Symbolic, posited as governing principles, are capable of assimilating the object of the analytic experience? The notion of the Real, which remained in the dark for so long, will be used to this end; its task will be to simultaneously represent the content and the limits of this experience.

The Impossibility of the Real

The attempt to found psychoanalysis as a science thus leads us to elucidate the notion of the Real—and not by chance, as we shall see, since the origin of this Lacanian notion is to be found in the domain of the epistemology of science. To my knowledge, and contrary to what can be said for the Imaginary and the Symbolic, there is no text in which the Real as such is developed. The word appears here and there, as if its meaning were obvious. If it can be definitely stated that the Imaginary dates from 1936, and the Symbolic from 1953, it could not be said outright that the Real, given its name in 1953, actually came into its own from 1964 on.

As far back as 1936, in "Beyond the 'Reality Principle'," in the process of criticizing associationism, which he says seeks "a guarantee of truth," Lacan wants to introduce an "objective conception of psychical reality" (*E*, 74):

> In order to oppose it outright to a conception which is defined more or less judiciously in the theoretical foundations of diverse contemporary schools as the *function of the real,* let us say that associationist theory is dominated by the *function of the true*" (*E*, 75).

Further on, he stresses

> the ambiguity of a criticism which, on the basis of the thesis *nihil est in intellectu quod non prius fuerit in sensu,* reduces the action of the real to the point of contact with a mythical *pure sensation,* that is, to being merely knowledge's blind spot (*E*, 76).

To speak of the Real amounts to rejecting a fundamentally idealist theory in order to get back to an objective study of phenomena. The attitude of "submission to the Real in Freud" (*E*, 81) leads to a reliance on the subject's testimony:

> This is indeed the attitude common to a whole culture which guided the abstraction analyzed above as a scholarly abstraction: for the patient as well as the physician, psychology is the domain of the "Imaginary," in the sense of the illusory; what ever has a *real* meaning, consequently the symptom, can only be psychological "in appearance" (*E*, 81).

Taking the symptom seriously, therefore, means considering it not as something inconsistent, which passes as an illusion does, but as something resistant and solid, which the psychoanalyst will need to isolate:

> The work of an illusionist, we will be told, if it did not precisely have as its outcome dispelling an illusion. Therapeutic action, on the contrary, must be essentially defined as a double movement whereby the *image,* initially diffuse and fragmented, is regressively assimilated to the Real, in order to be progressively disassimilated from the Real, that is, restored to its own reality. An action which testifies to the efficacity of this reality (*E,* 85).

The image, as a formative Gestalt that has to be reconstituted, is not the Real, but it is of the order of the Real, from which it has to be disengaged. This Real has to be called efficacious, since the image allied with it determines and constitutes the subject, and explains the permanence subsisting throughout the incessant changes in appearance. Through the Real, thanks to the Real, the psychoanalytic object can be credited with the characteristics of objectivity.

If the term *Real* is still in flux here, making do, for instance, with a close proximity to reality, it nonetheless remains the case that what is real at this stage is the Imaginary.

It is not by chance, as has been noted already, that Meyerson's name happens to be mentioned in this 1936 text (*E,* 86), and that some subheadings, notably, clearly allude to his book *La déduction relativiste.*[17] The concept of the Real (then current in the work of philosophers like L. Brunschvicg and E. Le Roy) is used repeatedly in the book, and several chapters are even devoted to it. The main thesis of Meyerson's book, which an interpretation of Lacan cannot overlook, is that, in affirming the identity of being [l'être] beyond the hold that we are able to have on it, science is not unrelated to philosophy and common sense. Beyond the multiplicity of our perceptions of an object, we assume the existence of a real object independent of such perceptions. Similarly, contrary to positivism, it has to be conceded that science grasps not just relations, but objects. There is "a perfect similarity between the objects created by science and those whose perception, by a spontaneous act, posits their existence." The Real is what is situated outside us, it is the substrate of phenomena; it lies behind appearances and is independent of our consciousness; the Real is being [l'être], which underlies seeming [le paraître]. Science distances itself from anthropomorphic considerations—that is, "from considerations where the person of the observer intervenes, in other words, from what is referred to the self."[18]

Science thus overlaps with philosophy, since permanence, affirmed by science and independent of consciousness, assumes being: "The Real of relativist theory is, very definitely, an ontological absolute, a true being-in-itself, more absolute, more ontological still than the things of common sense and pre-Einsteinian physics."[19] Relativist theory "has as its goal informing us of the nature of this Real."[20] Even if the methods of deduction are not the same in mathematics and logic, both Hegel and Einstein rely on the same presupposition, namely that the Real is rational.[21]

Even if Lacan submitted the notion of the Real to major transformations, which will need to be spelled out, it nonetheless remains the case that for him it needed no explanation, since it was so tied to a commonplace of the philosophy of his youth. It was a word that formed a part of the language of the most rudimentary culture. Moreover, perhaps Lacan never abandoned the principal traits, which at the time made up the comprehension of the concept: The Real is an invariant that subsists and resists; it is independent of the self and of consciousness, it is the being of all phenomena, and finally, it is rational, which is why it is calculable and amenable to logic.

After 1936, the Real disappears from the scene for a long time. In 1951, in "Intervention sur le transfert" ["Intervention on the Transference"] (*E,* 226) and in 1953, in "Function and Field of Speech and Language,"[22] the Hegelian formulation whereby "everything which is real is rational (and inversely)" (*E,* 226) is taken up without further explanation. In the lecture given some months earlier, on "The Symbolic, the Imaginary and the Real," the latter is not developed at all. To my knowledge, it is only at the end of 1955, in the seminar devoted to *The Psychoses,* that the notion of the Real again assumes a notable place, in describing the phenomenon of hallucination. Lacan turns his back on his earlier conception linking the Real to the Imaginary, since it was from the Imaginary that the structure underlying psychical activity had to be extracted. It is always the same Real, at least in part, that he refers to, the one that underlies and accounts for reality.

The classical conception, which has it that "in psychosis, the unconscious is on the surface, is conscious" (*P*, 220) must initially be considered as being without interest, because the unconscious does not derive its efficacy from the fact that it is not conscious, but from the fact that it is language. "The question is not so much one of knowing why the unconscious which is there, articulated on the surface, remains excluded for the subject, non-assumed—but rather why it appears in the Real" (*P*, 20). The question is not given an explicit answer, but it is discernible throughout the text, and can be formulated as follows: If the unconscious appears in the Real, it is because it *is* the Real, because it is identifiable with the Symbolic.

Lacan avoids such a conclusion, since it would lead him back to what he sees as the classical position: In psychosis, the unconscious becomes conscious—but also because, in order to found his science, he needs a Real that is distinct from the Symbolic.

In fact, the classical position is somewhat different: What becomes conscious in hallucination in the form of visual and auditory images is the unconscious, inasmuch as it is linked to the Imaginary, and thus to the Imaginary no longer understood as solely specular, but as a source of images and the locus of forces and drives. In other words, according to the classical position, it is the Imaginary that appears in the Real, and which is then taken for reality by the psychotic.

For Lacan, this is inadmissible, since for him "translating Freud (. . .), the unconscious is a language" (*P*, 20). But for the time being, the following conclusion is unavoidable: If the unconscious becomes conscious for the psychotic, this amounts to saying that the Symbolic moves into the Real (here meaning reality in visible form). This leads to the formulation: "Everything which is refused in the Symbolic order, in the sense of *Verwerfung*, reappears in the Real" (*P*, 21). In order to account for this claim, Lacan refers to Freud:

> As you know, we are concerned with the Wolf Man, who is not without certain psychotic tendencies and properties, as he demon-

> strates in the short bout of paranoia he goes through between the end of his treatment with Freud and the time that he once again comes under observation. Well, the fact that he rejected all accession to castration, though it is still apparent in his behavior, into the register of the symbolic function, and that any assumption of castration by an *I* has for him become impossible, is very closely linked to the fact that he happened to experience, in childhood, a brief hallucination which he reports in extremely precise detail.
>
> The scene unfolds as follows: While playing with his knife, he had cut his finger, which was only hanging on by a thin shred of skin. The subject recounts this episode in a style adhering closely to lived experience [le vécu]. All temporal markers have seemingly disappeared. He is subsequently sitting on a bench, alongside his nurse, who is normally his confidant for his early experiences, yet he dared not tell her about it. This suspension of all possibility of speaking is terribly significant—and to the very person in whom he habitually confided, especially things of this kind. What we have here is a gulf, a temporal abyss, a break in experience, in the wake of which it emerges that there's nothing wrong with him, it's all over, and let's not talk about it any more. The relation that Freud establishes between this phenomenon and this very special *not knowing anything about it, even in the sense of the repressed* expressed in his text, can be translated thus—what is refused in the Symbolic order, re-emerges in the Real (*P,* 21–22).

Lacan's reasoning seems to be as follows: The Wolf Man's inability to speak meant that the threat of castration could not be metabolized, symbolized, and it reappears in the form of an hallucination—that is, in the form of a visual image, which for the psychotic has all the appearances of a real image.

As it stands, this would be comprehensible, but it is not what Lacan actually says. He implies that by confiding in his nurse the patient could have avoided the hallucination; but the hallucination took place before he had the opportunity to talk about it. More bizarre still, in his general formulation "Whatever is refused in the Symbolic order reappears in the Real," Lacan adopts the psychotic's point of view, since this Real is reality only for the psychotic. In fact, "There is nothing at all wrong with [the child]": He confused the Imaginary of his hallucination with the Real of reality, a confusion that is simply retranscribed by those who are then accused of understanding nothing:

> The real handling of the object relation within the framework of an analytic relation conceived as a dual relation, is founded on the misrecognition [méconnaissance] of the autonomy of the Symbolic order, which automatically brings with it a confusion of the Imaginary plane and the plane of the Real (*P*, 23).

Thus, those bad psychoanalysts who handle the object relation by confusing the Imaginary (in its current conception) with the Real of reality, have done nothing but describe what it is possible to understand in the psychotic's experience in order to distinguish it from the experience of the neurotic.

But Lacan does not see it this way. His adoption of the psychotic's point of view—that is, moving from a description of hallucination that applies exclusively to the psychotic to a general formulation that applies to everyone—was made possible, or even necessary, by his conception of the Imaginary. Indeed, since the Imaginary (reduced to the specular) cannot be the source of images that appear in hallucination (in this case, images of castration), they will be attributed, after having been specially selected,[23] to the Symbolic.

But adopting the more general formulation has a further advantage: It allows Lacan to suggest the existence of a link between the Symbolic and the Real. If this Real were simply reality as hallucinated by the psychotic, it could not be of any assistance in an elucidation of common experience. But if you retain the ambiguity of the word "real," things are different: What the psychotic *perceives* as real (that is, as what is reality for him) is what we have to take to be the Real (the substrate of apparent reality; in this case, the threat of castration). Here Lacan simply rehearses what he had already implied in identifying psychic causality with the causality of madness:[24] The psychotic is the one who reveals the Real to us—real causes, real entities and, specifically when he renders the Symbolic visible, the cause of all causes, things in the Real, in visible and audible reality. For the psychotic, there is no difference between reality and the Real, since what he renders visible for us is the Symbolic. Thus, the question "why does the psychotic cause the Symbolic to appear in the Real?," now finds an answer: It is because he is unaided and unencumbered by the self and the Imaginary, because he is paradoxically in direct contact with the Symbolic, which he does not want to know anything about, which, for that very reason, is for him inescapable; it becomes the only reality there is, and which, since it is not external reality, is the constructed real.[25]

So there is nothing surprising in the fact that, some weeks later, Lacan declares that the Other who founds speech is "an absolute Other, beyond everything you can come to know," "beyond that reality" that we are (*P*, 62); this is "the Other as real" (*P*, 78). But as convenient as this shortcut is, it cannot make us overlook just how unstable the meaning of the word "Real" is, as evidenced, for instance, by the following passage:

> Once introduced into the play of symbols, your behavior is always rule-governed. In other words, when a marionette speaks, it is not the actual marionette that speaks, but someone behind the scenes. The question is one of determining what the function of the character encountered on this occasion is. What we [illegible] say is that, for the subject, it is manifestly something real speaking. Our patient doesn't say that someone else is speaking for her; the words [illegible] gets from it are her own, but not inverted, her own words are in the other she herself is, the little other [le petit autre], her mirror image, her counterpart [semblable]. "Sow" [truie] is given tit for tat, and you no longer know which comes first, the tit or the tat. That speech is expressed in the Real means that it is expressed in the marionette. The Other at issue in this situation is not located beyond the partner, it is beyond the subject itself—this is the structure of allusion: it points to itself in a beyond of what it says (*P*, 63).

"Sow"[26] is the word heard by a delirious woman interviewed by Lacan. According to him, "what is important here is that the word 'Sow' has actually been heard, in the Real." "Since there is hallucination, it is reality speaking" (*P*, 62). But which reality? The reality of objects? No, says Lacan, the Other who speaks is "beyond this reality" (*P*, 62). From this hallucinated reality referred to the Other, his remarks slide incautiously toward the conditions of recognition and the reciprocity of human relations. Then we come to the passage on the marionette cited just now. Once again, a link is therefore established between the psychotic's Real and the Other, the Real present in all speech, in other words the Symbolic.

Thus we find ourselves confronted with three kinds of Real. The first is that of the psychotic, who is convinced that what he hears is real and moves us to believe that his hallucinated reality is the substrate of reality (what is an adjective for him becomes a substantive for us). The second is the Real of the Other behind the scenes, beyond reality, to which has to be added, as we shall see, the Real of what is actually said. This third kind of Real is obviously not distinct from external reality, but it is interesting to note that Lacan calls on this third definition only when he needs to distinguish between the three registers of the Symbolic, the Imaginary, and the Real. In other words, when he wants to posit the Real as a third term distinct in its relation to the other two, he is forced to identify it with reality. It obviously has to be this way, since the three registers have to be differentiated, but the obviousness of it has to go unnoticed, otherwise, being the same as external reality, the Real would no longer have any theoretical interest.

The first move is to affirm the value of the trilogy. In the course of the same session of his seminar, Lacan states that "Concrete discourse is real language," "The signifying material is the Symbolic," and finally, "there is no doubt that signification is imaginary in nature" (*P*, 65). And a little further on, he says: "And then there is the Real, the truly real articulation, where the magician's pea has changed places.[27] Real speech, I mean speech insofar as it is articulated, appears elsewhere in the field, and not just anywhere, but in the other, the marionette *qua* element of the external world" (*P*, 67). The distinction therefore seems to have been clearly established, and all the more so since the three registers are again differentiated in the following session: "You will recall that, while remaining fully within the phenomenon of speech, we are able to integrate the three levels of the Symbolic, represented by the signifier; the Imaginary, represented by signification; and the Real, which is actual discourse in its diachronic dimension" (*P*, 75–76).

The second move is to efface from articulated language its links with external reality in order to assimilate it to the Real, the substrate of reality—meaning the Real that Meyerson discusses, the Real of science and philosophy. The passage continues: "The subject disposes of a whole signifying material—his language, whether native or not—and he uses it to introduce meanings into the Real" (*P,* 76). If this Real simply designated discourse, all discourse, as the preceding sentences led us to believe, it would have no scientific consistency. It is therefore necessary to slide surreptitiously from this discourse to the discourse of science. The interwoven equivocations in the pages that follow allow him to move from the third meaning of the word "Real" (the Real of discourse in general) to the second (the Real of the Other). These pages need to be cited in full:

> The notion of discourse is fundamental. Even for what we call objectivity, the world objectified by science, discourse is essential, since the world of science, which we always lose sight of, is before all else communicable—it is embodied in scientific communications. Even if you succeeded in performing the most sensational experiment, if it cannot be repeated by someone else on the basis of your account of it, it is pointless. It is this criterion that determines whether or not something is scientifically accepted.
>
> When I presented you with the table with its three different points of entry, I localized the different relations in which we are able to analyze the discourse of delirium. This graph is not the graph of the world, it is the fundamental condition of any and all relations. On the vertical axis, we have the register of the subject, of speech and the order of alterity as such, the order of the Other. The fulcrum of the function of speech is the subjectivity of the Other, that is, the fact that the Other is essentially the one who is able, like the subject, to convince and to lie. When I said to you that the sector of objects which are quite real has to be located in this Other, it is understood that this introduction of reality is always a function of speech. In order for anything at all to be assigned, in relation to the subject and to the Other, some foundation in the Real, there has to be something somewhere which does not deceive us. The dialectical correlate of the fundamental structure which makes deception a possibility in subject to subject speech is that there is also something which is *not* deceitful.
>
> Note well that this function is fulfilled in very diverse ways, depending on the cultural climate in which the eternal function of speech comes to function. You would be wrong to believe that it is

the same elements, similarly qualified, which have always fulfilled this function.

Take Aristotle: everything he tells us is perfectly communicable, and nevertheless the location of the nondeceptive element is essentially different for him and for us. And where is this element located for us?

Well, whatever those minds who cling to appearances (which is often the case with powerful minds, even the most positivistic amongst you, indeed, even those free of all religious notions) might think about it, the simple fact that you are living at this precise point in the evolution of human thought does not exempt you from what is rigorously and overtly formulated in Descartes' meditation, from God insofar as he is incapable of deceiving us.

This is so true that a personage as lucid as Einstein, when it came to handling the symbolic order that was his, referred to it—*God,* he said, *is malgn, but he is honest.* The notion that the Real, as delicate as it might be to penetrate, is unable to play the villain with us, and will not intentionally take us in, is, although no one actually dwells on it, essential to the constitution of the world of science.

That said, I concede that the reference to the nondeceitful God, the sole accepted principle, is itself founded on results obtained by science. Indeed, we have never ascertained just who reveals the existence of a deceitful demon in the depths of nature. But this is not to say that an act of faith was unnecessary in the first steps taken in science, and in the constitution of experimental science. We take it for granted that matter is not given to trickery, that it does not deliberately sabotage our experiments and blow up our machines. These things do happen, but only because we deceive *ourselves;* there is no question of matter deceiving us. This step, though, is not at all completely straightforward: for it to be taken so confidently requires nothing less than the Judeo-Christian tradition.

. . .

This decisive step, which consists in positing the existence of something absolutely nondeceitful (for which the expression "act of faith" is not entirely out of place), was made possible by the radical nature of Judeo-Christian thought. And it is essential that this step be reduced to this act. We need only reflect on what would happen if there were one particle too many in atomic mechanics, if we were suddenly to notice that not only are there protons, mesons, etc., but also another element which we had not counted on, a character capable of lying. At that point, no one would be laughing any more.

> For Aristotle, things are completely different. What is it, in nature, that makes him so sure of the nondeceitful Other, the Other as Real, if not entities inasmuch as they always return to the same place, namely the celestial spheres? (*P*, 76–78).

What to retain from this luxuriant text in order to elucidate the notion of the Real? We need to stick closely to the successive slidings and to what is motivating them. Initially, the Real was discourse—that is, actual sequences of words and sentences in which (symbolic) signifiers acquire (imaginary) meanings. But, since this discursive Real is nothing other than (sensible and audible) external reality, and is therefore at the Antipodes of the Real as substrate, it is reduced to the discourse of science. The latter definitely belongs to the order of the Real as Meyerson defined it, since it simultaneously assumes a rationality of theoretical principles and a corresponding rationality of the world of objects. For Lacan, the Real simultaneously covers the nondeceitful God, who is necessary to the constitution of all discourse, and matter, which is not given to trickery. Finally, the Other, who at the beginning of the passage was in need of a foundation in the Real, becomes the Other *qua* Real.

The Real, whose two aspects are the nondeceitful God and matter in science, therefore refers to the rational and to the Hegelian formulation that Lacan cites several times (*E,* 226; 310). He must have already read it somewhere in Meyerson's work. Here it has to be emphasized that the Real works perfectly well; if machines break down, "it is because we deceive *ourselves,* there is no question of matter deceiving us." Whatever doesn't go as planned is to be attributed to the part we play in it, it is not the Real's affair, which, because it is rational, is not deceitful. We are therefore a long way from the impossible Real that will make its appearance later on, but for now it is important for the Real to be presented in the guise of the Real of science.

In these pages, Lacan appears as the great unifier. He generously reconciles the God of the philosophers and the scientists with the God of Abraham, Isaac, and Jacob in the Judeo-Christian tradition. But he also makes use of what he has learned from the philosophers of science, namely the correspondence between mathematical formulae and the functioning of the material world. He even goes so far as to suggest that these remarks lead him to his reading of the Memoirs of President Schreber (*P*, 78).

But what is really going on? To begin with, the triad Symbolic/Imaginary/Real, which he wants to pass off as self-evident, has in no way been convincingly established. We saw that articulated discourse is not the Real, but quite simply reality. Moreover, even if all articulated discourse were assimilated to *scientific* discourse, it would still not be the Real. Indeed, no scientist confuses the Real with the mathematical formulae he works out; these formulae account for the Real, which is assumed to conform to them. The confusion of the two can only arise from the assumption that the substrate of external reality is in fact the unconscious as a language. At this stage, then, for Lacan the Real is still nothing other than the Symbolic. For, as we saw previously, it is the Symbolic that forms and informs reality. We are therefore forced to conclude that we are in fact not dealing with three terms, but with two: The Imaginary and the Symbolic, where the latter is *called* Real when it is posited as foundational.

Of the three meanings of the word "Real" mentioned above, the Real of the psychotic, the Other, and of articulated language, the third therefore has to be excluded, since it is related not to the the Real, but to reality, and is "there" only in the same way that a false window is "there." The second can be considered as the Real, but on condition of its being identified with the Symbolic; this leaves us with the first definition, which is referred to the situation of the psychotic, convinced that that what he hallucinates is real, and moving us to believe that this Real actually structures reality. Yet, once again, all we know of this Real is what we know of the Symbolic, since, according to Lacan's formula, what is refused in the Symbolic reappears in the Real.

But what does the psychotic contribute to the debate? You could say—the essential. For he is the only one who meets no resistance from reality; he is the only one who makes reality the product of his anxieties, anxieties that are characteristic of the human being. The psychotic is the one who brings what shapes and informs our existence into the domain of the visible and the audible. These are not just his own personal fantasies on display; these are the characteristic traits or fundamental laws defining every human being. This is essentially what Lacan sets out to suggest when he discusses the Wolf Man and the threat of castration that dogs him, castration having been posited, here and elsewhere, as the very crux of the Symbolic.

From within the chosen perspective of founding a science, the psychotic appears as the only true scientist. If you say that "for [the psychotic], the entire Symbolic is real" (*E*, 392), you are forced to conclude that he is the one who truly reveals the human psyche. Just as the physicist unveils the real of nature through his mathematical formulae, so the psychotic, through his attacks of delirium and hallucinations, brings to light the constitutive relations of speaking beings—in other words, the Real of human reality in general.

From another standpoint, the psychotic will also be the one who makes possible one of the most famous Lacanian adages. To confirm this, we need only go back to the formulations whose conjunction will lie at the origin of his subsequent determination of the notion of the Real. First, "Everything which is refused in the Symbolic order (. . .) reappears in the Real." Then,

> The fact that he expels all accession of castration—which nevertheless remains apparent in his behavior—into the register of the Symbolic, and that any assumption of castration by an *I* has for him become impossible, is very closely linked to the fact of his discovering that, in his childhood, he had experienced a brief hallucination, which he reports in extremely precise detail (*P*, 21).

If the two sentences are shortened and brought together, this is the result: Everything that is refused in the Symbolic order, since all assumption of castration by an *I* has for him become impossible, reappears in the Real. Or, more succinctly: What cannot be symbolized reappears in the Real. Or again: The Real is constituted by what it is impossible to symbolize. And, if "impossible" is changed from an adjective into a substantive, the result is: The Real is the impossible—meaning, implicitly, the impossible-to-symbolize.

Since this formulation has become, over time, a virtually universal definition, it has to be admitted that what is true for the psychotic applies universally, and that the psychotic is once again the model that allows an essential truth, a truth proper to all human beings, to be unveiled. But it needs to be stressed that, with the Real having become the impossible, we are suddenly at the very Antipodes of science. Here, nothing works any more. Could it be that matter has suddenly become deceitful [trompeuse]?

A different approach to the same formulation had already been presented the year before, in 1954, in the "Réponse au commentaire de Jean Hyppolite sur la 'Verneinung' de Freud," and also in relation to *Verwerfung,* at the time translated by the word *retranchement* (*E,* 386), one of the characteristics of psychosis.[28] After emphasizing that "what has not come to light in the Symbolic appears in the Real," Lacan comments: "For this is how the *Einbeziehung ins Ich,* introduction into the subject, and the *Ausstossung aus dem Ich,* expulsion from the subject, have to be understood. It is the latter [*Ausstossung*] which constitutes the Real in so far as it is the domain of whatever subsists outside symbolization. And that is why castration, here cut off by the subject from the very limits of the possible, but thereby also withdrawn from the possibilities of speech, will appear in the Real" (*E,* 388). Even if the word "impossible" is not actually used here (merely the expressions "limits of the possible" or "the non-possibility of speech"), the later formulations are nonetheless in the wind. Symbolization has not taken place, but what appears in the Real is undoubtedly the Symbolic. Without the psychotic, then, there would be no constitution of the Real *qua* locus of the Symbolic in reality.

We are confronted with a striking paradox. The category of the Real is produced in order to explain the situation of the psychotic, or, in other words, the Real appears because the psychotic is unable to symbolize. Thus, the category of the Real applies exclusively to the psychotic. But the category also has to be universalizable—that is, it must remain valid for those who do succeed in symbolizing the Symbolic, those who accomplish the introduction into the ego. From paradox, then, we move into a contradiction that can be stated quite simply: A category that was invented for the psychotic because of his inability to symbolize, to speak the Symbolic, is then applied to all those for whom it *is* possible to symbolize.

But why this generalization? In other words, what is it in the

Lacanian system that makes it both possible and necessary? It is Lacan's conception of madness, referred to earlier, that makes it possible: Psychical causality in general is nothing other than the causality of madness, which reveals the nature of the human psyche. Obviously, the value of this causality could be questioned, as could this kind of identification, but to the psychiatrist-psychoanalyst, they seem to present no difficulty. And second, the generalization is made necessary, since, if psychoanalysis wants to become a science in the manner of physics, as is implied in the long citation from the seminar on the *Psychoses,* it needs to find a Real for itself; and for the moment, the psychotic is the only one capable of providing one.

The contradiction then takes on new proportions. At least periodically in 1954, the purpose of the psychoanalytic cure is still seen as retrieving the blanks in the patient's history, so that he might assume it "in so far as it [his history] is constituted by speech addressed to the other" (*E,* 257); in short, so that the patient learns to symbolize. But this initial view has to be rejected, since if the Symbolic is appropriated, it will no longer appear in the Real, and so there will be no science. The whole science of psychoanalysis, the science of the Real, actually rests on the inverse of what is pursued in the cure, namely on the inability to symbolize, for this alone is constitutive of the Real.[29]

Far from pausing when things are going so well, Lacan throws himself into the project even more intrepidly.[30] To facilitate the reading of what follows, the following guiding thread may be proposed: The famous adage is going to find itself turned around—that is, it is going to shift from the subjective to the objective; the impossible will no longer have to do with man, but with things. The psychotic produced the Real because of his incapacity to symbolize; from now on, however, the Real will become whatever resists symbolization. The two "Reals" in question obviously no longer have anything to do with each other, since the Real of the psychotic is a creation that mimics the Symbolic, whereas the new Real, proposed in order to explain certain characteristics of neurotics or of human beings in general, is an obstacle, an insuperable limit, a dead end. In the case of the psychotic, it was still possible to think in terms of a structuring substrate; in the second case, we are faced with something totally obscure, known only through the force of its impact. But, since the same formula—the Real is the impossible (to symbolize)—can be used in both cases, it will be possible to suggest that we are actually dealing with one and the same entity.

We are therefore inclined or at least tempted to think that the Real in question no longer has anything to do with the Real of science, for it is hard to see how this Real-become-impossible could conceivably be the object of a mathematical interpretation. But, as Lacan used to say when, on the basis of his remarks, we happened to reach what to us seemed an inevitable conclusion: That is precisely what you *musn't* think. On the contrary, then, in this seemingly desperate situation, we are supposed to see ourselves as being at the very heart of mathematics and logic. Quite simply—but you really had to think about it, you had to have the audacity to think it and to say it—the impossibility of mathematicizing and logicizing the object of psychoanalysis actually reveals the very essence of mathematics and psychoanalysis.

These last remarks are in anticipation of the developments to follow, but they do sum up quite well the movement of Lacan's thought in his effort to make the notion of the Real consistent. We saw that, given his initial premises, his efforts could only end in failure, a failure he recognizes and which he simply turns into a cornerstone of the edifice.

In 1964, in the seminar entitled *The Four Fundamental Concepts of Psychoanalysis,* he declares that "No praxis is more oriented than psychoanalysis is towards that which, at the heart of experience, is the nucleus of the Real" (*Q,* 53). For him, this Real always remains the substrate of appearance, what lies behind and beyond appearances, since he borrows "from Aristotle, who uses it in his search for cause" the word *tuché,* which Lacan translates as "encounter with the Real" (*Q,* 53).[31] Indeed, he goes on to say that "The Real is beyond the *automaton,* beyond the return, the coming-back, beyond the insistence of those signs which we see governing us through the pleasure principle. The Real is what is always lurking behind the *automaton,* and it is perfectly obvious that, throughout Freud's research, this is his main concern" (*Q,* 53–54). In his analysis of the Wolf Man, Freud still clings to "asking himself what the first encounter is, the Real, which can be said to lie behind the phantasy" (*Q,* 54). "What cause is" (referring to Aristotle), "what lies behind," "what lies beyond": All these expressions clearly show that Lacan has not abandoned his project to isolate a real foundation. But it is at this point that his thought pivots: In accordance with what happens in the transference, he sees the relation to the Real as a relation of absence. More than that, like the trauma, it will be considered a failed or missed relation [relation manquée]: "The function of the *tuché,* of the Real as encounter—the encounter in so far as it may fail or be missed [la rencontre en tant qu'elle peut être manquée], in so far as it is essentially the missed encounter—first presented itself in the history of psychoanalysis in a form that was in itself already enough to attract our attention, that of the trauma" (*Q,* 54). The Real therefore presents itself "in the form of what there is in it of the *unassimilable*" (*Q,* 55). Lacan finds several examples of the trauma that provokes repetition—whether it be, for a father, the irreparable damage done by the death of his son (*Q,* 58), or, for a child, its mother's first ever absence (*Q,* 61), or the deep anxiety in a particular dream that gives it its sense of destiny (*Q,* 66). These examples are not chosen haphazardly, since their task is to prepare the audience for a new definition of the Real, the Real as lack, as failure, as a gap and a hole.

In the course of these seminars, the notion of the drive is introduced periodically, in anticipation of developments still to come. The drive indeed presents a major obstacle to Lacanian doctrine. We know that Freud makes it the basis of the unconscious, and that, for him, the drive is a force or an energetic charge which has "its source in a bodily stimulus."[32] But Lacan does not want to hear any talk of force or energy; he therefore has to propose a different interpretation, and to do so, he begins by situating the drive as an extension of the trauma:

> The place of the Real, which stretches from the trauma to the phantasy—in so far as the phantasy is never anything more than the screen concealing something quite primary, something determinant in the function of repetition—this is what we now have to locate. (. . .) The Real may be represented by the accident, by the slight noise, the small element of reality which is evidence that we are not dreaming. But, from another perspective, this reality is not so slight,

> for what wakes us is the other reality hidden behind the lack of that which takes the place of representation—this, Freud tells us, is the *Trieb* (Q, 58–59).

The difficulty presented by this remainder, the certain something that does not enter into repetition and therefore does not fit the logic of the signifier, is seemingly met head on. But a strategy emerges that will turn the problem of the drive around, first by introducing some lengthy remarks on the *object a* as the gaze, then by reducing the drive to a montage rotating around this object. The Real is then able to reappear in the form of the impossible (this time, the impossible-to-satisfy), in the form of the gap and the lack.

The interest of the invention of the *object a* lies in its being a hinge, a kind of crossover point for several different networks. The link between the trauma and this object is initially established through a commentary on the game that Freud's grandson plays with the cotton reel. The trauma consists in his mother's absence, to which the child responds by making the cotton reel, attached to some string, disappear and reappear; "[this real] is a small part of the subject that detaches itself from him, while still remaining his, still retained." Lacan takes this opportunity to explain that this object designates the subject in his relation to the signifier:

> If it is true that the signifier is the first mark of the subject, how can we fail to recognize here—from the simple fact that this game is accompanied by one of the first oppositions to appear [*fort-da*]—that it is in the object to which the opposition is applied in action, the cotton reel, that we must designate the subject. Later on, we shall give this object the name it bears in Lacanian algebra—the *petit a* (Q, 60).

Here the *object a* does indeed seem to be an effect of the signifier, or of the subject's submission to the Symbolic.

Somewhat further on, though, the *object a* is, on the contrary, aligned with the Real:

> It is here that I suggest that the interest the subject takes in his own split is tied to what determines it—namely, a privileged object, which has emerged from some primal separation, from some self-

> mutilation induced by the very approach of the Real, an object whose name, in our algebra, is the *object a* (*Q*, 78).

In this sentence the relation of cause to effect can be read in several ways: Either the splitting of the subject is determined by the *object a*, or the *object a* is produced by the subject's self-mutilation at the approach of the Real, or the Real induces the primal separation out of which the *object a* arises. It matters little which one comes first: With this lost object, which is an originally lost—indeed, a mythical—object, although Lacan sees it as an algebraic sign, the essential aim in these seminars is to establish it as an intermediary between the subject and the Real in order to pave the way for spiriting away the problem posed by the drive.

Further on, the *object a* becomes "something from which the subject, in order to constitute himself, has separated himself like an organ. This serves as a symbol of lack, that is, of the phallus, not as such, but in so far as it produces a lack" (*Q*, 95). And then, on the following page, he says: "The *object a* is most evanescent in its function of symbolizing the central lack of desire, which I have always indicated univocally by the algorithm (-phi)" (*Q*, 97). Here, then, the *object a* is aligned with lack, which is another name for the Real, but this new definition will allow it to play a role when the time comes to settle the question of the relations between the unconscious and sexuality.

In the course of several seminars devoted to the gaze, the model for the *object a*, the Real is no longer mentioned. It reappears—and not by chance, since it is still a question of founding a science—under the auspices of Newton, Einstein, and Planck. As we shall see, this obligatory patronage has now become somewhat strange. Lacan begins by recalling what he had formulated in 1953, but this time he goes a great deal further, and above all, sounds a new note, that of failure [ratage]:

> In my Rome report, I proceeded to a new alliance with the meaning of the Freudian discovery. The unconscious is the sum of the effects of speech on a subject, at the level where the subject is constituted out of the effects of the signifier. This makes it clear that, in the term *subject*—this is why I referred it back to its origin—I am not designating the living substratum required by the subjective phenomenon, nor any sort of substance, nor any knowledgeable being in his

> pathos [aucun être de la connaissance dans sa pathie], whether secondary or primordial, nor even an embodied *logos*, but rather the Cartesian subject, who appears once doubt is recognized as certainty—except that, in our approach, the bases of this subject prove to be considerably wider, but at the same time much more amenable to the certainty that eludes him. This is what the unconscious is.

And he goes on to say:

> There is a link between this field and the moment, Freud's moment, when it reveals itself. This is the link I express when I compare it with the approach of a Newton, an Einstein, a Planck, an a-cosmological approach, in the sense that all these fields are characterized by tracing in the Real a new furrow in relation to the knowledge that could be assigned to God for all eternity. Paradoxically, the difference which will most ensure the subsistence of the Freudian field is that the latter is a field which, by nature, goes astray and is lost [se perd]. And it is here that the presence of the psychoanalyst as witness to this loss is irreducible. (*Q*, 115–16).

This passage shows us that Lacan has certainly not abandoned his project to establish a science, and that he is still bent on situating it in the line of the philosophers who saw the Real as the locus of knowledge of the substrate and the rationality of phenomena. He will insist on this point by employing the word "cause," but in such a way that it is no longer clear just what knowledge he could have in mind, since this cause is lost, forbidden, impossible:

> This indicates that the cause of the unconscious—and you can see that the ambiguity of the word "cause" is to be taken seriously here: a cause to be supported, but also a function of the cause at the level of the unconscious—this cause must be conceived as, fundamentally, a lost cause. And this is the only chance we have of recovering it. That is why, in the misunderstood concept of repetition, I brought out the centrality of the perpetually avoided encounter, of the missed opportunity. The function of failure [ratage] lies at the center of analytic repetition. The meeting is always missed [manqué]—this is what constitutes, in comparison with *tuché*, the vanity of repetition, its constitutive occultation.

And then a little further on, we find this conclusion:

> At this point, I should define unconscious cause, neither as an existent nor as an *ouk on*, a non-existent—as I believe Henry Ey does, a non-existent of possibility. It is a *mè on* of the prohibition that brings an existent into being in spite of its non-advent, it is a function of the impossible on which a certainty is based (Q, 117).

This is not the first time that Lacan moves from fact to essence.[33] Using as his pretext the fact that, in the cure, it becomes patently obvious that words and sentences at times take on double meanings, he deduced that the subject was divided. Now, because the patient undergoing analysis finds it extremely difficult to abandon the repetition of his symptoms, we are supposed to conclude that the whole experience is doomed to failure. Moreover, he assumes that—and this is by no means obvious—this failure must be the result of a cause that the patient fails to attain, Lacan's conclusion then being that the cause is lost. This leads him to the final generalization: It is the Freudian field itself, which, by nature, tends to go astray. Thus, a perfectly rudimentary analytic experience, which could readily be more modestly explained, takes on the grandiose demeanor of tragedy and metaphysics, without neglecting to bestow on it the label "scientific." Guignol proceeds no differently when he sets out to frighten small children.[34]

Before returning to the question of the drive, and in order to pave the way for his response to it, Lacan confronts a preliminary difficulty, whose relatedness has no need of explanation: "The reality of the unconscious is—and this an untenable fact—sexual reality. At every opportunity, Freud defended it, if I may say so, tooth and nail. Why is it an untenable reality?" (Q, 138). Even though the question remains unanswered, it is stated clearly enough in the filigree of the text that the untenability of the claim for the sexual reality of the unconscious lies in its relatedness to the theory of the signifier. This is why, in the pages that follow, libido is reduced to desire, and desire itself reduced to the Real as impossible:

> I maintain that it is at the level of the analysis—if we can take a few more steps forward—that the status of the nodal point by which the pulsation of the unconscious is linked to sexual reality must be

> revealed. This nodal point is called desire, and the theoretical elaboration that I have pursued in recent years tends to show you, by taking you step by step through clinical experience, how desire is situated in the dependence on demand—which, through being articulated in signifiers, leaves a metonymic remainder that runs beneath it, an element which remains undetermined, a condition both absolute and ungraspable, an element necessarily in deadlock, unsatisfied, impossible, misrecognized [méconnu], an element which goes by the name of desire. This is what forms the junction with the field defined by Freud as that of the sexual agency at the level of the primary process (Q, 141).

Insofar as it is expressed in and through demand, desire is related to language, and thus to signifiers, yet it cannot be fully expressed in and through demand; insofar as it is linked to sexual reality, it has all the hallmarks of the Lacanian Real: It is impossible, ungraspable, while remaining absolute—all of which had to be suggested in order to prepare for what follows, which will be to the effect that sexual reality is defined by adjectives (which will eventually become substantives), provided that they bear prefixes of negation.

Lacan is now able to approach the formidable problem presented by the drive. This notion haunts his seminar that year; he had been grappling with it from the outset (Q, 49, 59), but it took all of the preceding detours to try to successfully deal with it. Once again, he meets the objection head on:

> Whereas the [non-psychoanalytic] past of the term "unconscious" weighs on its use in analytic theory, by contrast, when it comes to the term *Trieb*, everyone uses it as a sort of radical datum of our experience. People sometimes tend to invoke it against my doctrine concerning the unconscious, which they see as an intellectualization—if they knew what I think of intelligence, they would surely reconsider this criticism—and see me as in some way neglecting what every analyst knows from experience, namely the domain of the drive. We will indeed meet with something in experience that has an irrepressible character even after being repressed—besides, if repression there must be, it is because there is something on the outside which is pushing its way in. You don't need to go far in an adult analysis; you need only be a child therapist to be familiar with the element that constitutes the clinical weight of each of the cases we have to handle,

> which goes by the name of the drive. There thus seems to be a reference here to some ultimate datum, to something archaic, primordial. Such a recourse, which, in order to understand the unconscious, my teaching invites you to abandon, here seems inevitable (Q, 147–48).

The objection is completely forthright, and seems to have been clearly recognized. How will Lacan dispense with it?—essentially, by dealing with satisfaction, which Freud says[35] is the aim of the drive: "The use of the function of the drive has for us no other purpose than to place in question the status of satisfaction" (Q, 151)—by which he plainly means that nonsatisfaction is the only correct way of talking about the drive. And sure enough, as examples of this satisfaction, we are presented with sublimation, where the drive is said to be "inhibited with respect to its aim," "it does not attain it" (Q, 151), and with symptoms, which have to do with satisfaction (Q, 151), but the very least that can be said about them is that they are not terribly satisfying, that they amount to a state of satisfaction that has to be rectified at the level of the drive (Q, 152). What Lacan had to suggest is that the drive is to be understood solely in terms of satisfaction, and that, as far as satisfaction goes, there simply isn't any: Satisfaction is impossible. This is why he goes on to say that:

> This satisfaction is paradoxical. When we look at it closely, we notice that something new comes into play—the category of the impossible. In the foundations of the Freudian conceptions, this category is an absolutely radical one. The path of the subject—to use the sole term in relation to which satisfaction may be situated—the path of the subject runs between two walls of the impossible. (. . .) The impossible is not necessarily the opposite of the possible, otherwise, since the opposite of the possible is certainly the Real, we will be led to define the Real as the impossible.
>
> Personally, I see nothing standing in the way of this, especially since, in Freud, it is in this form, namely as the obstacle to the pleasure principle, that the Real appears. The Real is the impact felt, it is the fact that things don't fall into place straightaway, that is, as the hand held out to external objects would like them to. But I think that this is an illusory and quite limited view of Freud's thought on this point. The Real is distinguished, as I said last time, by its separation from the field of the pleasure principle, by its desexualization,

> and by the fact that, consequently, its economy admits something new, which is precisely the impossible (*Q*, 152).

This is somewhat surprising, since at no time during the seminar did we see it established that the Real admitted the impossible into its economy, unless it is simply because, in the psychotic, we encountered an impossible-to-symbolize, and because the Real was presented as a missed encounter, an impossible-to-know, and the drive is now defined by the impossibility of satisfaction. Assuming that these various "impossibles" have more than just the word itself in common, we would obviously need to know more about the links between psychosis and the Real, between "not knowing" [inconnaissance] and the Real, as well as the links between the drive and the Real. There is no attempt here to elucidate the different terms being used; clearly, the important thing is that they are all focused on this contradictory Real, which comes to be both an obstacle and a void.

The final move is to reduce the drive to the *object a*. It was stated earlier that the drive encounters the impossibility of satisfaction. Therefore, since "the drive, snatching at its object, learns in a way that this is precisely not the way it is satisfied," since "no object can satisfy the drive," and since "the object of the drive is a matter of indifference," we are led to give the *object a* "its place in the satisfaction of the drive." This definitively lost object could be said to be the object of the drive, but then the drive would lose itself in it. Now since the drive is a constant force, it will be said that it tends toward its object by ceaselessly shying away from it, that it therefore "circles around it" (*Q*, 153). This point is developed at length in the following session of the seminar: The drive is a *montage* whose "aim is nothing other than this return in circuit" (*Q*, 163); its only function is to "sidestep the eternally missing object" (*Q*, 164).

By linking what he has just said with his theory of the signifier, Lacan is able to conclude that:

> This articulation [of the drive and the object] leads us to consider the manifestation of the drive as an acephalous subject, for everything is articulated in it in terms of tension, and has no relation to the subject other than one of topological community. I have been able to articu-

> late the unconscious for you as being situated in the gaps that the distribution of the signifying investments sets up in the subject, and which are represented in the algorithm in the form of a losange, which I place at the heart of every relation of the unconscious between reality and the subject. Well, it is to the extent that something in the apparatus of the body is structured in the same way, it is because of the topological unity of the gaps in play, that the drive assumes its role in the functioning of the unconscious" (Q, 165).

What are the possible consequences of this account of the drive?

1. Let us assume that we are actually dealing with topology here (even though we know that topology is based on the hypothesis of the continuous). The unconscious is initially defined by the intervals between signifiers; it exists in the gaps; it *is* a gap. But, in that case, the drive, with its closed circuit forming a hole, adds nothing: The "subject riddled with holes" [sujet troué] is posited prior to any reference to the drive. In the Lacanian system, the drive is therefore a useless entity.

2. And insofar as it circles the *object a,* forming a hole, it is equally useless, since this "eternally missing object" (Q, 164) is itself already a hole. The obvious conclusion is that Lacan was obliged to discuss the drive because it constitutes a centerpiece of Freud's work, and because Freud remains the reigning authority in the field. But Lacan's interpretation of the drive drains it of its meaning, so that it can easily be dispensed with.

3. For Freud, the libido, the drive, trauma, and desire had distinct and well-established functions; it was impossible to confuse them. But once Lacan reinterprets them, these notions lose their principle of differentiation: The "sexual color" of the libido becomes "the color of a void: suspended in the glow of a gap" (*E,* 851); the drive has the "closed structure" (Q, 165) of a hole or a gap; the trauma is "inassimilable" (Q, 55), desire is "in deadlock, unsatisfied, impossible, misrecognized" (Q, 141).

4. For lack, holes, gaps, the impossible, and the Real to be true concepts, it would have first been necessary to give them precise definitions that precluded their interchangeability, and they would then need to have been placed in a fixed relation to each of the corresponding

Freudian concepts. The passion for the negative that underlies all these metaphors inevitably leads to a generalized indistinctness.

The sexual reality of the unconscious has thus been reduced to the drive; the drive was subsequently interpreted as a *montage,* and this *montage* turned out to form a gap. But there is one drive that cannot be reduced to partial drives, namely the genital drive—which leads straight to the question of love and sexual difference. The question here is whether there might be some chance of encountering sexuality in a form other than that of lack—no such luck. Indeed, on the one hand "the genital drive is subject to the circulation of the Oedipus complex, and to elementary and other structures of kinship" (Q, 173), in other words, subject to the Symbolic whose presence divides the subject and sends him in search of the lost object. On the other hand, sexual difference is referred to sexed reproduction, which separates the living being from sexuality: "It is the libido as pure instinct for life, that is, for immortal life, irrepressible life, for a life which needs no organ, a simplified and indestructible life. This is precisely what the living being is deprived of by virtue of the fact that he is subject to the cycle of sexed reproduction" (Q, 180).

This is a strange form of reasoning, which, parenthetically, reintroduces the biological, even though in the course of the seminar Lacan repeatedly states that the biological has nothing to do with the drive. For, even if it is true that the living, sexed individual dies after having reproduced himself, he is still a bearer of life, linked with living. Something of immortal life is conferred on him; he is not stripped of his libido. Lacan, however, finds it absolutely essential to retain nothing of sexuality apart from lures, gaps, lack, and therefore death. Later on, we even discover that "the drive, the partial drive, is fundamentally the death drive, and in itself plays the part of death in the living sexed being" (Q, 187). He has gone as far as he possibly can in emptying Freud's conception of the drive of every energetic and dynamic aspect.

And the reasoning is doubly strange, since, if sexed reproduction is introduced in order to claim that living beings are separated from it, it is hard to see why he wouldn't refer to it when discussing the genital drive. Indeed, biological genitality brings out the clear difference between the sexes. But no, the biological argument is to be employed when it authorizes stripping the living being of his libido, and rejected when the psychological argument leads us to think there is no tangible difference between men and women. This allows Lacan to achieve his aim, and to again indulge in his strange form of reasoning:

> Sexuality is established in the field of the subject by way of lack. Two lacks overlap here. The first one has to do with the central defect around which the dialectic of the advent of the subject to his own being in the relation to the Other revolves—by the fact that the subject is dependent on the signifier and the signifier first of all belongs in the field of the Other. This lack comes to take up the other lack, which is the real, earlier lack, to be situated at the advent of the living being, that is, at sexed reproduction. The real lack is what the living being loses, of that part of himself *qua* living being, in reproducing himself sexually. This lack is real because it relates to something real, namely this: that the living being, through being subject to sex, has fallen under the sway of individual death (*Q*, 186).

The inevitable conclusion is that sexuality is simply inadmissible in Lacanian doctrine, which is why it is reduced to a hole. There is no way in for it through the Symbolic, as we have just seen (we saw "life" meet with the same fate in the seminar on *The Psychoses*); nor could it find its way in through the Real, since, in the end, the Real can only represent the impossibilities and the empty spaces in the Symbolic. If, in Lacanian doctrine, sexuality is defined exclusively with reference to lack, it is quite simply because sexuality is lacking in Lacan's doctrine. It is all very well for Lacan to periodically defend himself against the objection that "I neglect the dynamic, which is so evident in our experience—and it is even said that I end up overlooking the principle affirmed in Freudian doctrine that this dynamic is, in its essence, sexual through and through" (*Q*, 185); whatever one might think of the Freudian doctrine itself, this objection remains unanswered throughout the seminar. Once you have managed to familiarize yourself with the subtle detours and infinite complications of Lacan's text, your only conclusion can be that the sexual reality of the unconscious has vanished into thin air.

Far from giving up and turning back, though, Lacan will keep going even more determinedly in the same direction, all in the name of the science that he so obstinately wants to found. The question of sexual difference, which, as we have just seen, had been outlined in the seminar of 1964, is central to the seminar *Encore* (1972–1973). Its main theme can be stated as follows: The sexual relation cannot be written down in mathematical form, which proves that psychoanalytic discourse is scientific—a singular claim, which certainly requires some explanation, and which will lead us back to the heart of Lacan's attempt to construct a science of the Real.

He takes as his point of departure the fact that love, although one fondly dreams of achieving it, is unable to turn two into one [faire de l'un avec deux]. This time the Real figures as a remainder, as something impossible to attain. The *jouissance* of the other always entails a limit, as can be illustrated by Zeno's paradox:

> Achilles and the tortoise: such is the structure of pleasure [le schème du jouir] for one side of the sexed being. Once Achilles has made his move, once he's shot his wad with Briseis, the latter, like the tortoise, has made a little progress, because she is *not whole,* not wholly his [*pas toute,* pas toute à lui]. There is a remainder. And then Achilles has to make his second move, and so on. That's even how it is these days, only now we've managed to define the number, the true, or rather, the Real. Because what Zeno overlooked is that the tortoise is not exempt from the fate that hangs over Achilles—its own steps also become shorter and shorter, and it will never reach the limit either. And it is on this basis that any number, if it is real, is defined. A number has a limit, and to this extent it is infinite. Clearly, Achilles cannot overtake the tortoise, he cannot even catch up with it. Only at infinity does he catch up with it. (*Eo,* 13).

The next page needs to be discussed in its entirety, because it is an excellent summary of the thesis developed throughout the remainder of the seminar. To begin with, the only *jouissance* that we know anything about is phallic *jouissance*, and because this latter is marked by a limit, there is a remainder that lies outside it:

> And there you have the last word on *jouissance,* inasmuch as it is sexual: on the one hand, *jouissance* is marked by this hole which leaves it no path other than that of phallic *jouissance.* On the other

> hand, might something be reached which would tell us how that which is so far merely a fault, a gap in *jouissance,* might be realized? (*Eo,* 14)
>
> It is because *jouissance* is phallic that there cannot be anything of the One in it, that this [phallic] *jouissance,* which is always missing something of the woman, makes a hole, a fault, a gap appear in this *jouissance.* No doubt, if we were to let ourselves get carried away with our dreams of fusion, if we were to play the angel [si nous faisions l'ange], we might discover this "One." But, since this is not the case in reality, we can only play the fool,[36] after the fashion of the parrot in love with Picasso:
>
> Curiously enough, this is what can be suggested only by some very strange insights. "Strange" [étrange] is a word that can be broken down into *the being-angel* [l'être-ange]—this is certainly something that the alternative of being as stupid as the parrot mentioned just now puts us on our guard against. Nevertheless, let us look closely at what is suggested to us by the idea that, in the body's *jouissance,* sexual *jouissance* has the privilege of being specified by an impasse (*Eo,* 14).

If we don't want to be considered fools, we should thus reflect on the fact that phallic *jouissance* creates a fault, a gap, an impasse. This seemingly negative fact will soon be turned around and bring us within reach of the pinnacle of contemporary mathematics, namely, topology:

> In this space of *jouissance,* to take something restricted, closed, is a locus, and to discuss it is a topology. In a work that you will see appear in the forefront of my discourse from last year [he is referring to "L'Etourdit"—"The Scatterbrain"—which appeared in *Scilicet* 4], I believe I demonstrate the strict equivalence of topology and structure. If we take our lead from that, what distinguishes anonymity from what we speak of as *jouissance,* namely what the law prescribes, is a geometry. A geometry is the heterogeneity of the locus, meaning that there is a locus of the Other. What do the most recent developments in topology allow us to propose concerning this locus of the Other, concerning a sex as Other, as the absolute Other? (*Eo,* 14).

Not only is the impasse transformed into mathematical positivity through recourse to topology, but it also allows the other sex, the female sex, to be posited as the absolute Other, a move that will later make it possible to situate in it a form of *jouissance* that is no longer exclusively phallic:

> At this point I shall propose the term "compactness" [compacité]. There is nothing more compact than a fault, provided it is quite clear that, with the intersection of everything in it which closes itself off being generally accepted as existing for an infinite number of sets, it follows that the intersection implies this infinite number, which is the very definition of compactness (*Eo*, 14).

Indeed, all the various "impossibles," all the various limits, all the impasses end up running into this hole, this absolute Other; they all intersect in this locus, which thereby acquires a matchless consistency. Psychoanalytic discourse thus becomes the foundation of all other discourses because their limits come to intersect in it, and because it is posited and supported in affirming the impossibility of the sexual relation. Psychoanalytic discourse is therefore the locus of all the limits and impasses that lie at the very heart of each and every one of these other discourses:

> The intersection in question is the one that I proposed just now as being what covers, what stands in the way of the supposed sexual relation.
>
> And only "supposed," since I state that analytic discourse is based solely on the statement that there is no—that it is impossible to posit—the sexual relation. There we have the advance of analytic discourse, and on the basis of which it determines the real status of all other discourses (*Eo*, 14).

This page is truly exemplary of Lacan's methods:

1. Supposing that (an opinion that can finally be attributed to Freud) phallic *jouissance* is the only form of *jouissance* that can be pinned down, why should *jouissance* in general be marked by a hole? Why couldn't it simply be said that there are other forms of *jouissance* that we don't know much about, or, if we follow Freud, forms derived from this primary *jouissance*? The task would then be to describe them. In Lacan's work, there is a kind of mental impotence when it comes to thinking the partial, the limited and the relative. If something is not whole, then it is marked, determined, produced, and caused by lack; but

these lacks, faults, and gaps then become entities in their own right, which, since they are not metaphysical, must be deemed mythical.

2. In this instance, what is it that makes this hole stand out for Lacan? He makes it perfectly explicit: Love, the dream of turning two into one. The gap simply consists in taking seriously the mad desire for fusion called love. Without it, the idea that *jouissance* suffers from a lack, that it is meant to lead to a single Being [l'Un], would never have occurred to him. The gap or impasse in question thus has the same nature as love—it is merely the frustrated version of it, and has therefore to be entered in the register of the fantasies that haunt the human imagination.

3. Lacan claims to elude substantification by resorting to topology:

> That this topology should converge with our experience to the point of allowing us to articulate it, is that not something which might justify that which, in what I propose, is supported (or worse, is moaned about) [se s'oupire],[37] by never resorting to any substance, by never being referred to being, by breaking away from whatever presents itself as philosophy? (*Eo*, 16).

But that would be assuming two things: On the one hand, the topology would need to have avoided relying on a prior substantification of gaps, faults, and impasse. But, for the hole to be able to gain admittance to the topology, it has to have been previously substantified on the basis of clinical experience; nowhere here is there anything to suggest that it is deduced from any kind of topology. On the other hand, a true topology would need to be established with respect to *jouissance,* phallic or otherwise. Later on, we shall discover that we don't even have the beginnings of one, meaning that, throughout, Lacan is doing nothing but turning fantasies into substances; in other words, he is producing myths, and whether their hue is topological or geometrical changes nothing in their nature.

4. The end result of all these operations can only be tautology. Take, for example, the sentence cited just now: "The intersection I speak of is the one I proposed just now as being what covers, what stands in the way of the supposed sexual relation." If we recall that this intersection is merely the compactness of the fault, then the intersection and

the fault are one and the same. But this fault was, in turn, discussed only in order to translate the impossibility of the sexual relation seeking to fuse two entities into one. One can only agree with Lacan on this point, but he cannot be said to have made terribly much progress in his thinking. The manifold detours were supposed to lead us to believe that analytic discourse determines the status of all other discourses, yet they have merely led us back to where we started. They have shown us, above all, that Lacanian psychoanalytic discourse, as a result of inflating itself with pseudomathematics and an outstripping [dépassement] of philosophy, has simply ended up exploding in our faces. All it has revealed, in fact, is its own gaps—in other words, its own emptiness.

Lacan's starting point was the impossibility of the sexual relation, which simply meant that the union lovers dream of is never realized. The same point is taken up again later, and aligned with writing and the relationship between science and writing:

> If there were no analytic discourse, you would go on talking like birdbrains, go on chirping the current discourse, go on playing the same record, the record that goes on playing because *there is no sexual relation*—and there we have a formula which can only be articulated thanks to the whole construction of analytic discourse, which I've been drumming into you for a long time now.[38]
>
> But after drumming it into you [vous la seriner], I still have to explain it—it is supported solely by writing [l'écrit] in that the sexual relation cannot be written down. Everything that is written stems from the fact that it will always be impossible to write the sexual relation as such. And it is from there that a certain effect of discourse called writing [écriture] derives (*Eo*, 35–36).

Lacanian analytic discourse is to be situated in the movement of molecular biology (*Eo*, 43), of Bourbaki (*Eo*, 31), or of Kepler (*Eo*, 43). The real break in science supposedly comes with the usage of alphabetic writing: "This is what tears us away from the imaginary—yet still founded in the Real—function of the revolution" (*Eo*, 43). Lacan immediately goes on to say that "What is produced in the articulation of this new discourse, which emerges as the discourse of analysis, is that it takes as its point of departure the function of the signifier, which does not depend on the signified for its effects, as even lived experience itself shows." This means that the new analytic discourse is firmly situated in the overall movement of the exact sciences:

> Nothing seems to better constitute the horizon of analytic discourse than the use made of the letter in mathematics. The letter reveals in discourse that which, not by chance, not without good reason, is called grammar. Grammar is what is revealed of language only through writing. Beyond language, this effect, which arises as a result of being based exclusively on writing, is undoubtedly the mathematical ideal (*Eo,* 44).

In short, on the one hand Lacan is suggesting that psychoanalysis has attained the status of science through its use of writing and, on the other hand, that the sexual nonrelation, which had previously been a vague, constantly reiterated assertion, has now become a scientific formula. But something is definitely wrong here. We can overlook the attribution of grammatical status to the manipulation of letters in algebra, since the main sophistry lies elsewhere. How can a formula or, more accurately, an assertion (for this is no algebraic formula even though the whole context seeks to imply that it is) be called scientific when it excludes the object in question from the possibility of mathematicization, because its formula cannot be written down? Either that or the phrase "The sexual relation cannot be written down" (*Eo,* 35) would itself have to be considered a mathematical formula, but this is precisely what Lacan rules out, since mathematicization is intrinsically dependent on the written word.

To claim that psychoanalysis alone, having entered the domain of science, is capable of positing the impossibility of the sexual relation is triply untenable: First, because this impossibility has no need of science for it to be posited; second, because it is not this impossibility that poses a problem for Lacanian doctrine; and third, because, on its own, this one impossibility is incapable of settling the question of the scientificity of psychoanalysis.

1. Let us assume for a moment that the impossibility of writing down the sexual relation makes analytic discourse scientific. But then

why does Lacan himself present us with the equivalent of this impossibility in terms that have nothing to do with science?

> It is in this that it is important for us to realize what analytic discourse actually consists of, important for us not to misrecognize that which doubtless has only a limited place in it, namely the fact that in analytic discourse we discuss that which the verb "foutre" states perfectly. In analytic discourse we talk about "foutre"—a verb, in English, *to fuck*—and we say that it just doesn't work out.[39]

And he goes on to say:

> Indeed, what constitutes the basis of life is that, as far as relations between men and women are concerned (what we call "collectivity"), things just don't work out. Things don't work out, and everyone talks about it, and the greater part of our activity is devoted to stating it (*Eo,* 33–34).

The adage "There is no sexual relation" merely translates in an extreme and unilateral way something that forms part of the common experience of humanity, and in which the emergence of science has changed absolutely nothing. The supposedly erudite formulation adds nothing to our understanding, and one is truly hard pressed to see why it is necessary to engage in algebra and topology in order to understand the phenomenon in depth. There is, in fact, nothing to understand here that everyone hasn't always understood—and all the more so, since the assertion "things don't work out" [ça ne va pas], which translates into the sexual nonrelation, the mainstay of psychoanalytic discourse, is quite simply false. Lacan says as much himself: "This relation, this sexual relation, insofar as it doesn't work out, works out anyway" (*Eo,* 34); moreover, "it doesn't work out" "has only a limited place" (*Eo,* 33) in analytic discourse. This, when translated, means: First, alongside the sexual nonrelation, there *is* a sexual relation nonetheless and, second, the sexual nonrelation is not the sole basis of analytic discourse.

2. The basic problem is not the sexual nonrelation, but rather phallic *jouissance.* Once the phallus has been posited as "the signifier that has no signified, the signifier supported in man by phallic *jouissance,*" and it has been further posited that "the signifier is situated at the level

of the "*substance jouissante*" (*Eo*, 26), and that "the signifier is the cause of *jouissance*" (*Eo*, 27), then language itself has been linked to *jouissance*. In other words, only phallic *jouissance* can be spoken, and is even intrinsically linked to speaking. It follows that the other form of *jouissance*, that of the woman, cannot be spoken: "There is a *jouissance* proper to the woman, and about which she herself perhaps knows nothing, apart from the fact that she experiences it (. . .); the woman knows nothing about this *jouissance*" (*Eo*, 69)—all of which places women in the company of the mystics and even God (*Eo*, 70–71). But if Lacan is forced to go to such extremes, whereby the positions of analytic discourse with respect to women exactly match those of Christianity, it is quite simply as a result of having posited as a first principle the claim that "the unconscious is structured like a language" (*Eo*, 46–47). These premises necessitate his slyest contortions:

> *Jouissance*, then: how shall we express what should not be [expressed] with respect to it [ce qu'il ne faudrait pas à son propos], if not like this—if there *were* a form of *jouissance* other than phallic *jouissance*, it would not necessarily have to be that one [i.e., that of the woman] (*Eo*, 56).

And then a little further on, he says:

> If there were [a form of *jouissance*] other than phallic *jouissance*, but there isn't, apart from the one women don't breathe a word about, maybe because they're not familiar with it—the one that makes the woman not-whole. It is false that there is another form of *jouissance*, which doesn't prevent the other half of the statement from being true, namely it [the other form of *jouissance*] need not be that one [again: that of the woman] (*Eo*, 56).

Thus, in the first place, phallic *jouissance* is not the only form there is; second, if the *jouissance* of the woman is the one that is ruled out [celle qu'il ne faut pas], it is quite simply because the doctrine has excluded it in advance.

3. We have seen several times that the relationship between analytic discourse and science is based on the statement that there is no sexual relation, since the latter cannot be written down. Yet the possibility of alphabetic writing is in other respects held to be the very mark of

scientificity. Therefore, it is no longer at all clear in just what sense analytic discourse is supposed to have entered the field of science. Either that, or we would need to be able to affirm a thesis underlying the entire discourse, to the effect that writing enables us to take a decisive step towards the Real that cannot be written down. And this is exactly what we find here, in a more complicated but no less absurd form, where the middle terms have been carefully kept well away from each another:

> Since for us it is a question of considering language as that which functions in order to make up for the absence of the only part of the Real which is unable to form itself out of being, or to form its own being [qui ne puisse pas venir à se former de l'être], namely the sexual relation,—what support is there to be found in reading only letters? It is in the very play of mathematical writing that we have to find the point of orientation towards which to steer ourselves in order to extract from this practice, from this new social tie which is emerging and spreading in a remarkable way, i.e. analytic discourse, what can be extracted from it with respect to the function of language (*Eo*, 47).

What is to be extracted from analytic discourse by means of the written in order to make up for the absence of what cannot be written? Lacan would like to have us believe that, thanks to analytic discourse, something has changed:

> Up till now, in the field of knowledge nothing has been conceived which did not participate in the fantasy of inscribing the sexual relation (*Eo*, 76).

The new order would thus be defined by the conviction that such an inscription is impossible. But saying that this inscription is impossible, saying that it is impossible to turn two into One [l'Un à partir de deux], will not abolish the fantasy, and if it is required for knowledge to be produced, then there is little chance of it being abandoned. But isn't this fantasy of inscribing the sexual relation in fact a form of the fantasy subtending Lacan's entire *oeuvre,* namely that of successfully translating psychical reality into an algebra, through the play of mathematical writing? In the end, this unavowed failure would have to be expressed in a singular proposition of the form: The Real of sexuality and the sexual relation is the remainder that cannot be introduced into the scientific edifice, but that which must be taken for the scientific Real, since, without science, it could never be deemed nonscientificizable.

In any case, Lacan has long been well aware of the results of his operations. In his work, the two walls of the impossible go by the names of tautology and the absurd, but he doesn't abandon his goal; on the contrary, as we have seen, he puts on a real performance—acknowledging his failure, but by turning it into the first and last principle of his doctrine. He can thus have his cake and eat it too, pushing the pretensions of his enterprise into the realms of extravagance, and claiming that this lunacy is reason itself. Witness one of the rare texts written directly for publication, and which dates from the period of his seminar *Encore*.[40] It is a text of truly dazzling mastery and agility, trickery and mystification, a text in which, either irritated or paralyzed, you are sure to lose your way if you haven't learned his devious ways beforehand, a text impossible to analyze, but from which we can extract a few pearls for the subject at hand.

The main project is announced on the very first page: "I remind you that it is through logic that this discourse [psychoanalytic discourse] makes contact with the Real, only to encounter it as impossible, whereby [psychoanalytic] discourse raises logic to its highest power: the science, as I said, of the Real" (*Sci,* 5–6). Here he makes no attempt to conceal the ambition that still ran along more or less in the filigree of the seminar *Encore*. But the ambition is announced as a *fait accompli;* all he does here is remind us of it. What he says amounts to this: The science of the Real raises logic to its highest power, meaning that analytic discourse would be logic to the power of two. Next, this science "which makes contact with the Real only to encounter it as impossible," does not in fact encounter the Real, since the Real cannot be written; what we're dealing with here is, precisely, the missed encounter [rencontre manquée] of the seminar on the *Four Fundamental Concepts* and the exclusion of the sexual relation from the field of writing, as in the seminar *Encore*. It is an odd logic, then, that manages to exclude from its field the very entity that constituted its object, and which, as a result, cannot be developed as a logic.

Objection sustained. Highlighting the formula "There is no sexual relation," the text continues:

> This means that of relation (of relation "in general"), a linguistic statement of it is all there is, and that the only assurance the Real has of a relation comes from the limit demonstrated by the logical consequences of such a statement.
>
> Here there is an immediate limit, owing to the fact that, with a statement, "there isn't" [n'ya], anything [with which] to form a relation.
>
> This fact has no logical consequences, which is undeniable [qui n'est pas niable], but which no negation is enough to support: only the statement "nya" [there isn't] [is enough to support it] (*Sci,* 11).

No logical consequences—what that means is: No logic at all. All that remains is the statement that logic is powerless to write the sexual relation down in mathematical form. What we encounter here is in fact this very powerlessness as constitutive of the Real. It is because logic runs up against the nonexistence of the sexual relation that this relation then becomes part of the Real [ce rapport devient du réel]. Lacan's reasoning here can readily be seen as conforming to the familiar distinction between "reality" and "real." As long as no attempt is made to introduce the sexual relation into the domain of logic, it remains a fantasized reality; if, on the contrary, it is confronted with logic, the sexual relation will then be of a mathematical order, and will form part of this "furrow traced in the Real" (*Q,* 116), it will be wrested from the Imaginary and the fantasy of a single Being [le phantasme de l'Un]. But, for this to be the case, the sexual relation, which exists only fantasmatically, would at least have to be girded by logic, rather than by sheer assertion, but this is precisely what Lacan has just ruled out.

That the logic proposed by analytical discourse has no logical consequences is something that had already been stated as early as 1960:

> That is why, at the risk of falling out of favor, we have indicated how far we have been able to push the diversion [détournement] of the mathematical algorithm in our use of it: the symbol square root of −1, which is still written as 'i' in the theory of complex numbers,

> is obviously justified only because it makes no claim to automatism in its later use (*E*, 821).

This plainly means that nothing of a mathematical nature can be elicited from Lacanian mathematics and no logical reasoning emerges from it; it is sheer misappropriation [détournement] of mathematics. What we are presented with, then, are nothing but images, metaphors, and illustrations, as stated on several occasions—for example, concerning the Borromean knot (*Eo*, 115, 116), or, elsewhere, concerning irrational zero.[41]

If, therefore, the logic girding the sexual relation has nothing to do with logic (it hardly needs to be said—with logic *proper*), if it is no longer anything but an image, albeit with a mathematical tinge to it, then we are still firmly in the grip of the fantasy of one Being, simply expressed in a different way, meaning that it is this pseudoscience, founded on the sexual nonrelation, that founders.

That the text of "The Scatterbrain" should then launch into an elegy on impasse and equivocation as keys to the new science of the Real is scarcely surprising, for only impasse and equivocation could be signitaries to this enterprise. But can an elegy turn this desperate situation into a triumph? Is the impasse proof, by the mere fact of its existence, of the rigor of what preceded it? When the paradoxes of logical discussion "from before the time of Socrates" and in Cantor and Russel are invoked in order to justify equivocation, it is a case of witnesses being called before a bar that doesn't concern them. Indeed, in every single case, it is the various forms of mathematics themselves that, in the course of their development, encounter obstacles or aporias, although this is not the case here, where impasse is merely preceded by equivocation. Indeed, what kind of logic would it be that sought to base itself exclusively on obstacles and aporias? In fact, this is no longer logic, but merely a stale odor of negative mysticism that posits the apophantic as the basis of its knowledge—which has obviously become a nonknowledge [non-savoir]—and which then becomes all the more fascinating in not being governed by its procedures.

Because in the preceding discussion we have not gone into the arcana of the Lacanian topology, in its antecedents and its supposed consequences, we would do well to refer to some connaisseurs. A psychoanalyst and a mathematician have gone to considerable trouble, with the greatest respect and esteem, in order to study Lacan's most recent advances in these matters from a purely logical standpoint. On the basis of their conclusions, we shall see whether, after 1973, the Lacanian enterprise had any chance of success. This is what they tell us:

> Before being able to specifically approach the relations between the Symbolic and the topological—the interest of this latter being the possibility of it leading to the Real—it is important to reconsider the current conception of *Borromean knotting,* the topological organization that we are employing, of the three dimensions R.S.I.—the Real, the Symbolic and the Imaginary. Indeed, it is to Lacan that we owe both the introduction of these three fundamental categories whose usage in psychoanalysis is generally accepted, however divergent this usage might be, and the elucidation of the Borromean properties. But this second structure serves precisely to organize the first—they have to be considered jointly. It seemed necessary to us, in order to introduce our thesis, to suspend judgment on the proposition asserting the Borromean character of R.S.I., so as to, on the one hand, set the three domains *free*—and to take up the question of their respective or joint consistency—and, on the other hand, to isolate the Borromean nodal logic, thereby allowing the question of its interest to be posed. Nevertheless, this move has to be adequately supported.
>
> There is nothing to be lost by placing in question the hypothesis formulated by Lacan in 1973, and maintained since then: It is not *obvious* that the Real, the Symbolic and the Imaginary are knotted together, nor that this knotting takes a Borromean form: the proposition does not have an axiomatic status. For it to become obvious may well be thought to entail a lengthy effort, either in the field of science—of which we have given no account so far—or in the field of psychoanalysis. But the testimony of psychoanalysts is rather conflicting on this point, and this disparity, which doesn't prove anything, extends as far as the elementary definitions of what is at stake. Should this be held against psychoanalysis? Certainly not, if we consider that psychoanalysis in itself has nothing to do with the object or discourse of science. Over and above this, there is the fact that metaphorical usage—and not *metaphorical* usage of a *model,* which has an effective rationale in the order of science—of topological constructions and the consequences such usage eventually has on the *logic* of discourse, is perfectly valid. But Lacan went beyond

> this metaphoric usage when he introduced his *mathemes,* of which the Borromean knot is a cardinal form. Lacking axiomatic status, the proposition can be formulated as a postulate, that is, its promoter and those who follow him on this point *request* an adherence in principle, guaranteeing that the development that follows from this new foundation will demonstrate its validity through the extension of its rational field. So far we have made no progress in this direction, which leads to the observation that this postulate is, instead, a dogma, and that it is sustained solely through the faith invested in it.
>
> If there is no great practical or theoretical disadvantage in leaving the Borromean character of the fundamental triad to one side, there is, by contrast, an obstacle to conceding it without criticizing it. This obstacle is the *necessarily discontinuous consistency* of the Symbolic. Indeed, we assume, in the absence of a radical change in definition, that any system of symbols is discontinuous, and that it is in the very *nature* of a sign to exist only through discontinuity, that is, through *distinctive opposition:* you do not go from one symbol to another by means of an infinitesimal transition. This discontinuity is an objection in principle to the Symbolic entering *in itself* into any form of topology. Indeed, we know that the *power of the continuous,* or the *continuum,* is an absolute condition for the constitution of a topology and, therefore, of a logical nodality.
>
> It is precisely the attention that we have given to this difficulty which has led us to pose the question of the logical conditions which would authorize an hypothesis on a possible liaison between symbolic organization and the logical continuum. The Borromean chain seems no doubt partially capable of supporting this function. But, as has been said, we are not going beyond a reading-writing, the reading being organized at the level of the Symbolic, whereas the writing refers to topological figuration. It has to be emphasized that we have not, therefore, reached a solution which would make it possible for R.S.I. to form a knot, this time supported by proofs. The Symbolic remains fundamentally disconnected from the Real in the specificity of its functioning.[42]

In the course of their study, the authors are led to conclude that, from a logical standpoint, it is impossible to distinguish between the Symbolic and the Real:

> Each mode of functioning, consistency, or property of each domain corresponds term for term with that of the other: their organization

> is *identical* and, as a result, essentially indistinguishable (. . .) Everything that characterizes the so-called Real characterizes the so-called Symbolic.[43]

If I allow myself to cite this text at length, it is because it remains unpublished, and because, by situating itself in the register of mathematical rigor, it puts an end to any hope of founding a valid topology on the basis of the Borromean knot—as if, after 1973, Lacan had found the solution to the problems he had raised! For that to have been the case, he would need to have changed course, which never happened; the Real was subsequently always conceived in the same way. We saw that the developments concerning this notion, beginning in 1955 with the psychoses, had taken on, as of 1964, a more significant role. The Real had initially been posited as the impossible-to-symbolize, then subsequently appeared as the ungraspable aspect of the trauma, then as the impossible-to-satisfy of the drive, and finally, in 1972–1973, as the impossible-to-formalize of the sexual relation. Far from him breaking out of the close confines of these purely negative aspects of the Real, we find even more extreme formulations—for example, in the seminar of 1976, where he states that "The Real is to be sought on the other side, in the direction of absolute zero. (. . .) The bottom limit is the only real thing there is." "Its stigmata, the stigmata of the Real as such, lie in its not being connected to anything." Further on, we again encounter the luminous cloud of the Judeo-Christian tradition: "Darkness is only a metaphor here, because, if we had the merest scrap of the Real, we would realize that light is no darker than shadows, and inversely." And then comes a further invocation of death:

> *Trieb,* which is translated in French—and I don't know why—as *la pulsion;* the word 'dérive' [drift] could well have been used. The death drive is the Real in so far as it can only be thought as impossible, that is, every time it appears right under your nose, it is unthinkable. There is no hope of approaching this impossible, since this unthinkable is death, whose Real basis is that it cannot be thought.[44]

The escalation just keeps going in the same direction. By thinking the Real as the impossible, the ungraspable, the unassimilable, and the unthinkable, it is no longer an obstacle to symbolization, satisfaction, and formalization; it is no longer even a flaw bordered by compactness [compacité]. It becomes absolute zero, no longer connected to anything at all. Its only remaining option is to drift toward death. Does this mean the kind of death that sages and mystics have in mind when discussing the human world? Definitely not, for the Real we are now dealing with is not an unthinkable that could conceivably flow back into the Symbolic and the Imaginary in order to start afresh in rethinking them; it is the end of the line with no turning back, a dead end of knowledge.

In sum, it would be extremely difficult to give the notion of the Real any consistency at all. In one sense, it has no conceptual content apart from the Symbolic, since the Real is defined by the impossibilities encountered by the Symbolic. As an obstacle to symbolization, it could have become the Symbolic's "other," a reality to which the Symbolic did not have access. The definition of the Real as the impossible made constant reference to the reality of the Symbolic, yet remained empty. The Real is indistinguishable from the Symbolic, whether it is aligned with numbers (the unconscious can be counted, the Real is founded on real numbers), which was already the case with the Symbolic, or becomes a gap, a flaw or a hole, since these terms are already implicit in the discontinuity of the Symbolic.

In another sense, the Real is merged and confused with the Imaginary, as we saw in the seminar on the *Four Fundamental Concepts,* where it was reduced to the fiction of the drive, and in the seminar *Encore,* where the single Being [l'Un] that lovers dream of was identified with the impossibility of formalizing the sexual relation. The Real is also imaginary every time one seeks to turn gaps or holes into self-contained realities [une réalité en soi], every time it is cut off from everything else.

The Real is also at times reduced to external reality—for example, the reality of actual discourse, the reality of the obstacles one runs up against, of the objects produced by scientific techniques,[45] or the reality of death, to the extent that one avoids turning it into a metaphysical, mystical, or mythical reality.

But in the end it could be said that the Real simply does not exist, since, in one of its final forms, it is presented as the sexual relation; and if the sexual does not exist, then neither does the Real. Or, it simply does not exist as a concept, since it can go on drifting in all directions.

In any case, it is impossible not to wonder why Lacan made such frequent use of the word, why he wanted to make it a centerpiece of his discourse. These questions will provide us with an opportunity to go back over the doctrine as a whole, and the consequences it had.

Notes

1. Meyerson uses the term *identification* once, in a marginal work, in order to designate the identity of an antecedent and its consequent. Reason seeks to explain a phenomenon "by first aligning the consequent with its antecedent, then even denies all diversity in the species," *La déduction relativiste,* Payot 1925, p. 307.
2. See *Identité et réalité,* second edition, Alcan, 1912.
3. Marcel Mauss, *Sociologie et anthropologie,* P.U.F., 1st ed. 1950, 8th ed. 1983. I am using the latter edition here.
4. This influence has already been noted. The title of the 1953 lecture, "The Symbolic, the Imaginary and the Real," "points to the revelatory effect that Levi-Strauss's early theories had on [Lacan]," Marcelle Marini, *Lacan,* op. cit., p. 51.
5. In 1966, in "La science et la vérité," he again throws the burden of proof entirely on Levi-Strauss. And, in order to respond to the latter's eventual objections concerning the legitimacy of the transposition into psychoanalysis, Lacan calls on "experience."
6. On this point, see the decisive article by Vincent Descombes, on which I am largely drawing, "L'équivoque du symbolique," in *Confrontation,* no. 3, Spring 1980, pp. 77–95.
7. A term which Lacan had until then scarcely used, above all as a substantive, whether because, in Freud, the notion seemed to him too vague, or because he could not integrate it into his own doctrine. Until 1953, the unconscious is linked, for Lacan, to imaginary structures, imagos, forms, and complexes. See *C,* pp. 24–25 and 44; *E,* p. 108. In any case, it is Levi-Strauss who made possible the Lacanian use of the notion of the unconscious, for he is the one who suggested the link between the unconscious and language on the one hand, and language and formalization on the other. Freud was subsequently reintroduced to the circuit and forced into compliance with the new conception.
8. "Language and society," in *Structural Anthropology.* This article is the only one to be cited in a note in "The function and field of speech and language in psychoanalysis," *Ecrits,* p. 285.

9. A passage rewritten in 1966. The 1953 text reads: "To be sure, this *rapprochement* of what passes for the most exact science with the science which admits being the most conjectural seems surprising at first, but the contrast is not contradictory" (*La psychanalyse,* I, P.U.F., 1956, p. 131). There has been some progress made in the wording, but not in the proof.
10. We know that Levi-Strauss was in New York during the Second World War, where he worked with the linguist Roman Jakobson, and the mathematician André Weil of the Bourbaki group. This intersection of the three disciplines made structuralism possible. Cf. the Appendix to *The Elementary Structures of Kinship*.
11. Vincent Descombes, art. cit., pp. 81–82. Here I am merely summarizing the arguments presented in this article.
12. ibid., p. 82.
13. *Bulletin de l'association freudienne,* no. 1, pp. 4–5.
14. Lacan avoids using the word "subjectivity" here, but he will make use of it later, once he has given it a certain inflection.
15. On this confusion, which runs through Lacan's *oeuvre,* see Vincent Descombes, *Grammaire d'objets en tous genres,* éditions de Minuit, 1983, pp. 187–250.
16. In a recent book, by way of a commentary on the sentence: "For, barring a denial of what is the very essence of psychoanalysis, we have to make use of language as a guide in the study of what are called verbal structures," we find: "The brilliant reversal carried out is the following: if the analytic method is founded on speech, it is because language is the condition of the unconscious." This commentary goes a great deal further than Lacan's text, which remains extremely prudent. Written by a well-informed Lacanian, the commentary actually shows quite well what Lacan wanted us to believe, and why he was considered a genius: he succeeded in deducing the nature of the object being studied from the method used to study it.
17. Payot, 1925.
18. ibid., p. 28.
19. ibid., p. 79.
20. ibid., p. 80.
21. ibid., p. 198.
22. An initial ambiguity is already apparent here, since the expression "anything real," employed on the same page, refers to the Real as external reality.
23. This reduction of the Imaginary to the specular Imaginary, and therefore to representation and to language, is also at work in the surreptitious movement from an hallucination that is here clearly sensory to a verbal hallucination, which will subsequently lead to the exclusion of all other forms.
24. "Propos sur la causalité psychique" (1946), *E,* 151–93.
25. In 1954, in "Réponse au commentaire de Jean Hyppolite," the identity of the

Symbolic and the Real is already noted: "This is what explains, it seems, the insistence that the schizophrenic puts into reiterating this step. But in vain, since for him, the Symbolic in its entirety is real" (*E*, 392).

26. The word "sow" refers to the animal. [trans.]
27. "la muscade passée dans l'autre": the reference is to the well-known sleight-of-hand magic trick played with a pea placed under one of three cups, which are then shuffled in such a way as to make it seem as if the pea has bafflingly moved from one cup to another. [trans.]
28. The word is a nominalization of the verb *retrancher,* which means "to cut off (from)," as in a person being cut off from the world. According to Laplanche and Pontalis (1973), the standard French translation is now *forclusion,* a term that Lacan himself proposed; the standard English translation of Freud's term is *foreclosure.* [trans.]
29. Understandably, Lacan's notion of the Real has discouraged even the most patient and lucid minds; as Maurice Dayan observes, "You never know what can happen in a Lacanian discourse on the Real and reality" *Inconscient et réalité,* P.U.F., 1985, p. 81.
30. This process, which consists in outrunning the difficulty and thus becoming even more entangled in it, brings to mind the Professor of History Antoine Blondin discusses, who was unable to pronounce the word "Westphalia," so that when he came to writing his account of the Treaty, he was forced to reinvent the whole story.
31. The Greek word *tuché* has to do with fortune, with happenstance and chance events. [trans.]
32. J. Laplanche and J.B. Pontalis, *The Language of Psychoanalysis,* trans. D. Nicholson-Smith, New York, W.W. Norton, 1973, p. 214.
33. He is, however, not unaware of the illegitimacy of this procedure, since he reproaches Freud with it: "How can it be said, just like that, as Freud does, that exhibitionism is the contrary of voyeurism, or that masochism is the contrary of sadism? He proposes this analysis for purely grammatical reasons having to do with the inversion of subject and object, as if the grammatical subject and object were real functions. It is easy to show that this is not the case, and we need only to refer to our structure of language for this deduction to become impossible (*Q*, 154–55). If Lacan had denied *himself* this kind of deduction, what would remain of his *oeuvre?*
34. Guignol is a popular French puppet character, originating in Lyon in the early nineteenth century; somewhat akin to Punch of "Punch and Judy" fame, he is seen as the French descendant of Pulcinella. [trans.]
35. This passage from the seminar is a fine example of Jacques Lacan's return to Freud. His commentary has only a few verbal links with the text; in particular, Freud speaks at length about the satisfaction of the drive; in the case of sublimation, he in no way says that there is no satisfaction, but that satisfac-

tion is partial. The misinterpretation is so flagrant here that it is unthinkable that Lacan was not aware of it and did not want it that way. He nonetheless plays the part of the well-informed reader: "You should never read Freud without first pricking up your ears" (Q, 153)—ears no doubt pricked up/ trained ["dressées": a word which also means "trained"—Tr.] (so as) to make Freud say what Lacan can turn to account. Freud must have had considerable authority in the circles in which Lacan spoke for him to have felt the need, at least after 1953, to invoke it. It is amusing to learn that, in 1966, when he republished his "Discourse of Rome" in the *Ecrits,* he corrected his text here and there by adding Freud's name.

36. Although the phrase "faire la bête" would normally simply mean "acting the fool," where "bête" means "foolish" or "stupid," Lacan's subsequent reference to the parakeet reactivates the literal meaning of "bête," namely "a dumb animal." Lacan is alluding to one of Pascal's best-known *Pensées:* "L'homme n'est ni ange ni bête, et le malheur veut que qui veut faire l'ange fait la bête." Blaise Pascal, *Oeuvres complètes,* ed. J. Chevalier, Paris, Gallimard, 1954, p. 1170. [trans.]
37. The "verb" "se s'oupire" is a Lacanian improvisation; he is punning on the phrase "ou pire," meaning "or worse," and the verb "soupir(er)" (to sigh), and seems to be using it as an echo of the verb "se supporter" (which immediately precedes it: "dans ce que j'avance, se supporte, se s'oupire . . .", namely the notion of what is *tolerated* ("supporter" means "tolerate" as well as "support;" hence "what is tolerated, or worse, what is moaned (about) . . .". [trans.]
38. "S'il n'y avait pas de discours analytique, vous continueriez à parler comme des étourneaux, à chanter *le disque-ourcourant,* à faire tourner le disque, ce disque qui tourne parce qu'*il n'y a pas de rapport sexuel*—c'est là une formule qui ne peut s'articuler que grâce à toute la construction du discours analytique, et que depuis longtemps je vous serine." Here we are faced with a Lacan even more playful than usual, and the strict impossibility of *translating* a fairly complex network of double meanings. Throughout the paragraph, Lacan is playing with an ornithological isotopy: "les étourneaux" is the word for "starlings," but is also a slang equivalent of "l'étourdit" ("scatterbrain"), hence "birdbrain;" "chanter" (to sing) obviously refers *back* to birdsong (hence "chirping"), and then *forward* to "le disque-ourcourant" ("un disque" being a record; "ourcourant" doesn't exist as a separate word in French) where Lacan is punning on "le discours courant" (current discourse); finally, the verb "seriner" means to teach a canary to sing ("un serin": a canary), but is also a slang term for harping on something. [trans.]
39. The French verb "foutre" (from the Latin "futuere," meaning "to have sexual relations with a woman"), is used in virtually the same—and as many—ways as "(to) fuck;" the common expression "c'est foutu" is close enough in mean-

ing to, if not as vulgar as, the English expression "it's fucked" ("ruined," "wrecked," "done for"). This seems to be what Lacan has in mind here by pointedly invoking the English word. [trans.]

40. "L'Etourdit," in *Scilicet,* no. 4, éditions du Seuil, 1973, pp. 5–52.
41. *Ornicar?,* nos. 26–27, p. 23. The proof that the Borromean knot is merely an image, a schema is provided by its usage: it does not always overlap with the trilogy RSI. Thus, on December 17, 1974 (*Ornicar?,* no. 2, p. 99), we find Lacan speaking in terms of inhibition, symptoms, anxiety or meaning, existence, and holes; whereas on the May 11, 1976 (*Ornicar?,* no. 11, p. 2), we find the Real, the Imaginary, the unconscious and the ego, and so on.
42. Denis Lécuru and Dominique Barataud, *Essai sur quelques raisons d'une lecture borroméenne du symbolique,* typewritten text, dépôt légal, 1982, pp. 10–11.
43. ibid., pp. 132–33.
44. Seminar of March 16, 1976, *Ornicar?,* no. 9, pp. 33–38.
45. Press conference given in Rome on October 29, 1974; published in *Lettres de l'École freudienne,* November 16, 1975.

3

The Principle of Incoherence

In the course of his seminar of January 11, 1977, Lacan claims to be "terrorized" after reading *Le Verbier de l'Homme aux Loups*,[1] a text that he sees as a continuation of his own discourse: "This 'Verbier,' what we have here, I think, is the furtherance of what I've always advocated, namely that the signifier is what the unconscious is all about." And he goes on:

> Psychoanalysis is not a science. It has no scientific status—it merely waits and hopes for it. Psychoanalysis is a delirium—a delirium which is expected to produce a science. We could be waiting a long time! There is no progress, and what you expect is not necessarily what you end up with.
>
> It is a scientific delirium, but this doesn't mean that analytic practice will ever produce a science. And this science has an even slimmer chance of reaching maturity since it is antinomic, and since, through practising it, we know that there are connections between science and logic."[2]

Lacan would therefore seem to be retracting his prior claims and assertions; he seems to be saying that psychoanalysis is not a science, and never will be. But what he actually says is much more complex, first, because psychoanalysis, which is not a science, is said to be a delirium "which is expected to produce a science," even if this is a vain expectation. Second, because this delirium has the adjective "scientific" bestowed on it, which proves that, as a delirium, psychoanalysis continues to maintain relations with science. And finally, because, even if the psychoanalytic science that emerges from delirium has little chance of reaching maturity, it regains scientific status through its relations with logic, as its practice shows us. Indeed, earlier in the seminar, Lacan had stated that "In the structure of the unconscious, grammar has to be eliminated. Not logic, but grammar."

On the one hand, Lacan seems to be breaking away from what he nevertheless recognizes as a consequence of his pronouncements:

> There is one thing that amazes me even more than the diffusion—which we know is taking place—of what is called my teaching or my ideas, in this thing which is on the move under the name of the Institute of Psychoanalysis,[3] and which is the other extreme of analytical groupings. What amazes me even more is that (. . .) one Jacques Derrida has written a fervent, enthusiastic preface to this "Verbier" (. . .) I find, I must say, in spite of the fact that I started things going in this direction, that neither the book nor the preface observes good form. As deliriums go, this is an extreme case. And it frightens me to feel more or less responsible for having opened the flood-gates.

But he keeps heading in the same direction nonetheless. His terrible fears do not lead to any form of retraction, and to confirm this, we need only juxtapose the statements that open and close the seminar. At the beginning he states that over the course of the previous 25 years he has always striven to speak the truth: "The truth about what? About knowledge [le savoir]. This is what I believed psychoanalysis could be founded on, since, in the final analysis, everything I've said holds together." And then at the end, referring to what he had said a little earlier, "The unconscious, in short, comes down to the fact that one speaks—if indeed there *is* any speaking-being [du parlêtre]—all by oneself [tout seul]," he concludes:

> It is in this respect that we speak to ourselves by ourselves [nous nous parlons tout seuls], to the point where what we call an ego emerges, with no guarantee against the possibility of its being, strictly speaking, delirious. It was in this sense that I pointed out that, like Freud for that matter, there is no need to look all that

> closely when it comes to psychoanalysis. Between madness and mental deficiency, all we have is a choice.

The definition of psychoanalysis as scientific delirium, therefore, covers not just the work of certain followers, but that of Lacan himself. If everything holds together as it does in madness (not mental deficiency, of course), psychoanalysis is a scientific delirium. It should not be thought that Lacan introduces these formulations just by chance, merely in order to impress his audience of the day, for there are others just like it. On November 24, 1975, at Yale University, Lacan had stated that "Psychosis is an essay in rigor. In this sense, I would say that I'm a psychotic. I'm a psychotic for the single reason that I've always tried to be rigorous" (*Sci,* 9). From rigorous psychosis, we have thus moved on to scientific delirium, which amounts to virtually the same thing. Lacan's discourse, then, is a discourse that has a certain consistency: It is scientific in the manner of a delirium, and rigorous, in the same way that psychotic discourse is rigorous. He feels there is nothing to worry about, since what we see as rigor in general is to be entered in the register of psychosis. Indeed, he goes on to say: "That's obviously going quite some way, since it assumes that logicians, for example, who tend towards this goal [rigor], as do geometers, would in the final analysis share a certain form of psychosis."

Obviously, it is hard to see why the fact that psychosis can appear to be an essay in rigor implies that logicians suffer from a certain form of psychosis. If we adhere to what he actually says here—and there is a certain prudence displayed in the wording—there is a veritable gulf between an essay in rigor and rigor itself, and there is definitely a further gulf between a certain form of psychosis and psychosis itself. And bridging these gulfs would require a great many explanations. But chances are that any clarification would end up destroying the thesis being advanced. Indeed, for Lacan, the important thing is to give the impression that if analysis, which is not a science, can be called a scientific delirium, it is because, inversely, rigor, which lies at the heart of logic, has to be seen as related to psychotic delirium. In this way, two terms indispensable to the solidity of the Lacanian edifice find themselves juxtaposed or even identified: Delirium and rigor, madness and logic. Indeed, we have repeatedly seen that the propositions advanced in the *Ecrits* and in the *Seminars* could be maintained only at the price of systematic confusion. In claiming that the delirium of psychoanalysis is scientific, and that rigor has to do with psychosis, Lacan is stating the fundamental principle on which his whole enterprise is based.

Once we can no longer make out the frontiers between science and delirium, logic and psychosis, between what holds together as delirium and what compels recognition as mathematical reasoning, we are no longer dealing with structurally well-defined paradoxes. We are plunged into an inability to establish elementary distinctions, an inability to show how and why any concept could be opposed to another. What has happened that, at the end of the road, the entire doctrine should acknowledge that a generalized state of dissolution is necessary for it to hold together? What relentless internal movement has been at work, for madness to have become the sole principle of cohesion?

The best way to answer these questions is to retrace our steps and try to determine whether, and when, a wrong road was taken, a road ending in impasse, forcing Lacan to plunge deeper and deeper into the absurd, and to brush aside even the most basic laws of language and communication.

As a result of our earlier analyses, we saw that the Real bore the marks of every negation and privation. These latter pertain, however, to two decidedly different registers. On the one hand, there are the negations concerning the eternally lost object, the lack that causes desire, the life excluded by sexed reproduction, and so on; and on the other hand, there are the negations engendered by the imperatives of the symbolization or writing of the sexual relation. From this we could deduce that the two kinds of negation, which are still called failures or missed encounters, have nothing to do with each other, since the first kind pertains to the tragic aspect of existence where one never succeeds in finding the wholeness one aspires to, whereas the second kind stems from the insistence on scientificity in psychoanalysis. This is undoubtedly correct, for, in Lacan's *oeuvre,* on the one hand there is a tone of despair and a passion for loss, which account for one of the inescapable aspects of the human condition; and on the other hand, we find an unshakeable audacity seeking to reintroduce into knowledge the most obscure aspects of this same condition.

But if you assume the standpoint of the doctrine, a single proposition is actually responsible for provoking the appearance of both forms of lack: "The unconscious is structured like a language." Indeed, this one proposition simultaneously excludes the object of research and includes the possibility of a logic founded on linguistics. Lacan summed up this situation clearly in his lecture of November 24, 1975, given at Yale University:

> What struck me when I read these three books [*The Interpretation of Dreams, The Psychopathology of Everyday Life, Jokes and their Relation to the Unconscious*], is that Freud's acquaintance with dreams was limited to the account which was given of them. It could be said that the real dream is ineffable and, in numerous cases, it is. How can the real experience of the dream *be* [comment peut être l'expérience réelle du rêve]? This was one of the objections raised against Freud: it lacks validity. For it is precisely on the material of the account itself—the way in which the dream is recounted—that Freud works. And if he carries out an interpretation, it is an interpretation of the repetition, the frequency, the weight of certain words. If I had a copy of *The Interpretation of Dreams* with me, I could open it at any page and you would find that it is always the account of the dream as such—as verbal material—which serves as the basis of the interpretation.
>
> In the *Psychopathology of Everyday Life,* we find exactly the same thing. If there were no account of the slip or the parapraxis, there would be no interpretation.
>
> The major example is provided by the joke, where the the quality and feeling of satisfaction displayed by the participants—Freud insists on this—come essentially from the linguistic material.
>
> That led me to state—what seems obvious to me—that the unconscious is structured like . . . (I said 'is structured like', which was perhaps exaggerating a little, since it presupposes the existence of a structure—but it is absolutely true that there is a structure) . . . the unconscious is structured like a language (*Sci,* 13).

Here again we encounter Lacan's quite characteristic sophistry: Since we know the unconscious—source of the dream, the slip, and the joke—only through language, the unconscious is structured like a language. Which is tantamount to saying: Since we can only know certain objects by looking at them, these objects are structured like eyes.[4] This sophistry will enable him to carry out a twofold operation, which would otherwise have been unthinkable: On the one hand, the object in question, precisely the Real, "the real dream," "the real experience," is going to find itself cast out into the ineffable, where it becomes lack; on the other hand, this same object, which now belongs to the order of language, can then be subjected to the procedures of knowledge [connaissance], and why not, while you're at it, to the ultimate requirements of scientific knowledge [connaissance]. In this way, the pathos surrounding the eternally lost object, lacks, gaps, holes, and death, can be married with the loftiest logical aspirations simply by means of a glaring blunder in reasoning.

The operation carried out on the dream will similarly serve to domesticate affect, life, and lived experience [le vécu]. In a lecture given in Milan in February, 1973, for instance, after noting that "after all, affect, apart from the fact that it's not intellectual, its definition is not readily grasped," Lacan gives a further turn to the difficulty or, more precisely, gets rid of it through recourse to language: "It is at the level of the object's [la chose] articulation with language, its articulation with a signifying support that, if I can put it this way, what used to be affect is secondarily justified" (*I*, 79–80). It is therefore quite clear that affect has no reality of its own; what reality it has is secondary, because—and to the extent that—it is articulated verbally. The following year, once again in Milan, he says the same thing about lived experience: "Since I don't know what life is, and I've stressed this before, I don't know what lived experience is either." "It is precisely at the level of the fact that it is spoken that one notices that it conceals something which did not appear at all, to begin with, in its lived aspect [dans son vécu]; it conceals a knowledge [un savoir], and this is what Freud designated under the name unconscious" (*I*, 113). We are thus still dealing with the same strange form of reasoning: Since we can know affects, life, or lived experience only to the extent that they are spoken, they therefore belong to the same species as speech.

Consequently, in its turn, and as was foreseeable, the theoretical instrument that emerged from all this sophistry conspires to miss the very target it had been designed to hit. Indeed, if "to recount a dream is something which has nothing to do with the dream itself, the dream as it is lived" (*I*, 112)—and the same can be said of affects—then it is not surprising that, as we saw in the seminar on *The Psychoses,* the Symbolic, that is, language (to the extent that, for Lacan, this is more specific) should prove to be incapable of assimilating life, singularity, and subjectivity. Nor is it surprising that the Real, which was supposed to be the object of knowledge, should have managed to slip through his fingers, and that in order to designate it Lacan should have had to resort to expressions that mark a powerlessness to grasp even the vaguest semblance of it.

In short, then, from the moment that the unconscious is posited as being structured like a language, and turned into a form of knowledge,[5] because he wanted to retain only the verbal manifestations of the unconscious—the verbal account of the dream, affects to the extent that they can be described, and lived experience to the extent that it can be interpreted, the question of their reality no longer even being considered—Lacan is able, and deliberately seeks, to overlook the fact that these phenomena have to do with a play of forces, that they have a reality and a dynamism of their own. What remains blurred in the definition of the unconscious as structured like a language (for we never know exactly what the words "structure" and "like" mean) disappears if it is considered purely and simply as knowledge, and even more so (an inevitable consequence) if man is defined as a "speaking-being" [parlêtre]. It's always the same form of sophistry at work: Under the pretext of speech being a distinctive characteristic of humanity, our attention is drawn to this characteristic to the exclusion of all others. The withering rejection of the mythology of the affective, ceaselessly denounced, can only engender a further mythology which is oblivious to itself, like, for example, the mythology of an unconscious defined by the letter.

Once again, if the Real is void, if it can be apprehended only negatively, it is quite simply because the theory of the Symbolic has, from the outset, barred access to it, and because it is by definition incapable of grasping anything of it. In declaring that language was the sole object of psychoanalysis, Lacan believed it could be given a scientific basis, for he was then on terrain where something objective could be grasped. But in reality, something quite different happened: The instrument psychoanalysis employs in order to function took the place of its object, an object that belongs—and this bears repeating—to the order of subjectivity, singularity, affectivity, and life. On the pretext of finally revealing the object of psychoanalysis, Lacan placed himself on familiar ground, but this meant forgetting what he was there to find out.[6] With the unconscious structured like a language, something is knowable, but there is nothing left to *be* known [on peut savoir quelque chose, mais il n'y a plus rien à savoir].

Lacan's failure to account for the Freudian unconscious does not, however, date from the introduction of the Symbolic—it goes back much further than that. Indeed, if, in the wake of Levi-Strauss, Lacan was able to forge his own conception of the Symbolic, it was because he had previously armed himself with a theory of the Imaginary, on the one hand viewed as a structure, and on the other, reduced to the specular (whence, as a consequence, the importance assigned to the gaze as *object a*).

We recall that, in order to account for human behavior, Lacan appealed to an underlying form, a substrate: Beneath the apparently infinite diversity of words, gestures, and actions, there lies a principle of cohesion, a self-image constituted by the relation to the other, by the image of the other with which the individual identifies in alienating himself. Whence a new understanding of aggressivity, now seen as the result of the necessity, and impossibility, of recapturing for oneself one's own image, an image forever captivated by the other. Lacan had thereby brought to light the function of the ego as the lure of mastery and as misrecognition [méconnaissance].

Furthermore, the Imaginary is exclusively defined in this context in its relation to the mirror image, and is reduced to this particular form, that is, to the specular. No mention is made of the Imaginary's ties with the imagination, which reveals itself in dream images, in the images of waking life, in artistic or poetic creation, in fantasy and the fantastic. This is certainly no accident, since these images slip through the fingers of objectivity and science because they are manifestations of an inalienable subjectivity, of the singularity of the individual, of his lived and felt existence, always finally incommunicable, which is why Lacan's scientific project had to ignore them as unassimilable.[7]

Now, all of these characteristics are to be found point for point in Lacan's conception of the Symbolic. The subject is alienated in the signifier, to the point of having no other existence than that of a pure locus of passage from one signifier to another. Similarly, the Symbolic refers to the structure subtending the subject, even if he goes as far as to disappear in it. The human individual had been reduced to the form that constituted him in imaginary rivalry; hence, at a later stage, we find that there are no subject-to-subject relations that are not governed by effects of signifiers.

Just as the structure defined by alienation in the image of the other meant the exclusion of the nonspecular Imaginary, so the language-like structure of the unconscious imperiously brings with it the exclusion from the unconscious of all other determinants, or their reduction to this structure, without which the project as a whole would have been threatened. Lacan, herald and champion of misrecognition, was therefore forced—because he wanted at the same time to become the unchallenged master of it—to found his enterprise on a misrecognition that made it possible. Reducing the imaginary to the specular, and the unconscious to a structure of language, is a remarkable operation which amounts to trying to bring into the world of representation whatever radically eludes it, so as to be able to ignore it.[8]

It remains to ask ourselves how it was *possible* for Lacan to develop his *oeuvre* over several decades, always heading in the same direction—that is, continually inventing new forms of equivocation, in an ongoing effort to construct an edifice that succeeded in capturing and holding our attention. There is, indeed, no doubting that what we have here is an *oeuvre*, and, through the sheer force of its incoherence, an *oeuvre* invested with a coherence of its own. For, if nothing in Lacan holds up to the most elementary logical scrutiny, if every page and sentence is a permanent challenge to the principle of identity and noncontradiction, it nonetheless remains the case that everything he produced bears the same indelible stamp. And, since we are not dealing with poetry, but with something that sees itself as theory, it means looking for a principle of unity somewhere, to which the name "the principle of incoherence" has to be given.

Lacan ceaselessly employs paradox (in the etymological sense of the word, that is, whatever goes against commonly accepted opinion or runs counter to good sense). But his thought is in no sense dialectical; there is no referring of a term to its opposite in order to bring out the link that exists between them. Rather, he constantly isolates one side of an opposition, and hurls the other side of it into oblivion. His thought is utterly unilateral.

In his description of what he calls the Imaginary, for example, the only aspect he retains is the alienation of the ego in the image of the other. Similarly, when he seems to be following Hegel's account of alienation, not once does he point out the equal importance assigned to appropriation. And in all of his commentaries on Freud's *Negation,* while he mentions expulsion out of the ego, he completely ignores introduction into the ego, which for Freud is its necessary corollary.

Many instances of this procedure could be adduced. Take the slogan: "The desire of man is the desire of the Other." No doubt this formulation brings out either the relation of man's desire to what is unconscious in it (as if the experience of desire coincided with the experience of being acted by an Other), or the importance of the desire of others in the determination of one's own desire. But, precisely, once this formula becomes the exhaustive definition of all desire, it excludes from the outset the possibility of appropriating desire. This is very much in evidence in the seminar entitled "Le désir et son interprétation," where Hamlet serves as an example in establishing the thesis, since "he himself doesn't meet up with his own desire, for he no longer has any desire, inasmuch as Ophelia has been rejected by him."[9] From this statement one might have concluded that, outside the extreme, the pathological case, encountering one's own desire is actually feasible. But, for the paradoxical formulation to remain intact, this is precisely what Lacan has to rule out.

The principle of incoherence thus spawned two figures of logic (or a-logic): Equivocation, which is encountered throughout, and unilaterality. The first allows for an assimilation of the most disparate elements: Through equivocation, a limitless communication between all

possible objects, all ideas, but also all disciplines, is established. It suffices to bring out a single common trait, whether at the level of vocabulary, or at the level of content, for pathways to open up in every direction. It is through this procedure that Lacan's *oeuvre* gives one the impression of having assembled the totality of human knowledge, and that psychoanalysis is raised to the status of the science of sciences. By virtue of equivocation, all inclusions are possible.

The second figure is the inverse: It comes under the heading of exclusion. This figure makes it possible for psychoanalysis to seal itself off in an infinite pretention, to deflect all objections, and even all possible questioning. If you haven't undergone the experience, you don't have the right to speak out, and, if you object to the experience, this just proves that you did not submit to it in the appropriate spirit. This same procedure facilitates the rejection of all the material unable to be assimilated by the theory, and allows the latter to retain the purity of an edifice that would be compromised by the infinite complexity and great obscurity of a simple description of experience. To equivocation's generalized sleight of hand is joined the resolute, overbearing assertion of the unilateral.

And there the principle of incoherence and its two by-products refer us directly to what we know of psychosis. A delirious patient ceaselessly assimilates whatever he encounters, since, for him, everything is a sign of how right he is, or of how well-founded his thought is, as well as an opportunity for him to add to it. At the same time, he is able to ignore even the most glaring evidence that might cast doubt on his construction; he is, above all, oblivious to the causes of delirium, and to the forces leading him astray.

As a result, the fact that Lacan was able to conceive psychoanalysis as a scientific delirium becomes intelligible. If, on the one hand, the whole *oeuvre* is propped up by a single principle, from which certain consistently applied rules are drawn, the *oeuvre* can be said to possess a certain rigor, which will lend it the appearance of a science. If, on the other hand, equivocation and the unilateral—its twin rules—are the only modes of reasoning (that is, absence of reasoning) governing it, the *oeuvre* will have taken on the most obvious characteristics of delirium. There was nothing haphazard about Lacan's description of his work as "critical paranoia," for the author of this *oeuvre* is not a psychotic; this is a discourse in a permanently controlled skid, even if it always—inevitably—maintains its own trajectory. But. like the psychotic, the Lacanian system is cut off from life, from affects, from subjectivity, and from all appropriation.

Thus, because Lacan constantly missed the object of psychoanalysis, he was forced to adopt, by way of an overall link, a style woven entirely of disconnection [déliaisons], and was driven to construct an edifice that became more and more extravagant. But, paradoxically, thanks to this crazy system, he was always able to give the impression of accounting for the most vital aspects of psychoanalysis and the Freudian discovery. In a way, through his style, he succeeded in mimicking what we think we know of the unconscious or, put more strictly, what we think we know of the form taken in analysis by the analysand's discourse. Equivocation is its stylistic hallmark, for there is then no juxtaposition or linguistic enticement—and they come from all directions—that may not prove fruitful. As for the unilateral, it underlies the revealing denegations and successive blindings, whose frontiers are constantly pushed back, but never actually crossed.

Here, however, there is an evident danger of a reversal. As Regnier Pirard wrote after devoting himself to an extremely close study of Lacan's use of linguistics, "This particular linguistics is more suited to mimicking the unconscious than teaching us what it means to speak. But does psychoanalysis consist in mimicking the unconscious, or in making it speak? And isn't a mimetic linguistics a superfluous and impotent redundancy? There is just a single step separating this redundancy from the emptying out of the unconscious. For if the unconscious invades language to the point of being equivalent to it, this amounts to saying that the unconscious no longer exists, that there has never been an unconscious."[10] In seeking to bring into conscious discourse all the supposed characteristics of the unconscious, you thus run the risk of suppressing the opposition between the conscious and the unconscious, and of thereby dissolving the conscious itself in a derangement modeled on your image of the unconscious. Paradoxically, once the unconscious is everywhere, you are radically cut off from it, which is precisely what happens to the psychotic.

Lacan was not the only one in France, in the 1960s, who dreamed of inventing a new form of rationality by giving ample space to madness, supposedly the only phenomenon able to reveal man in his essence. These people wanted to go much further than German Romanticism, which always maintained a certain distance between the relation to the world of dreams and fantasy as the necessary condition for creation, and the actual production of a work, which is subject to rules and assumes an order. The disorder of madness had to be allowed entry to writing itself, without allowing laws not conforming to madness—a madness supposed to be the sole bearer of man's truth—to intervene. Reason was seen as suffering from all kinds of ailments, and the repression it initiated had to be done away with, so that the great revelation might take place. This was a kind of hyperromanticism, which could well be another name for surrealism, for it was a question of paralyzing the rational as irrational, and thus allowing an unprecedented form of rationality to appear on the horizon.

There is a radical difference between trying to define the limits reason runs up against, and wanting to base reason on inviolable limits. In the first case, reason has its own domain, based on the principle of identity and the search for univocity; it recognizes that this domain has its frontiers, even if it strives to push them back. Reason is thus bordered by an exterior. In the second case, reason is no longer anything but an illusory superstructure, having no other justification than that of a system of defenses prejudicial to the emergence of truth. Since these limits are foundational, and are intrinsically necessary to the field they define, the field truly no longer has any limits; it is bounded by nothing except itself, is subject to no determination, and, *a fortiori,* no rules.

A further step is taken in this direction once the theory claims to base itself, not even on limits, but on impasses—that is, on taking wrong roads. Experience shows that all true research inevitably starts out by getting bogged down in successive dead ends before finding the judicious direction. In each case, it has to devote itself to sealing off false trails, in order to find others liable to lead to the desired goal. Establishing these impasses as foundational means condemning the entire field to sterility, and enclosing it—that is, if it were to accept any limits at all—in a sort of triumphant deadlock. A vainglorious desperation in which the clown who thinks he is pulling the strings is merely redoubling the madness he had posited at the outset as his first principle. The theory then takes on the form of a house built by Cheval, the postman: It may delight or fascinate the eye, but you could certainly never live and work in it.[11]

It no doubt required genius for Lacan to remain for so long at the crest of what he calls a scientific delirium. His friendly—and decisive—rivalry with Salvador Dali and Luis Bunuel, two great figures of the Surrealist movement, sustained him in the pursuit of his immense *oeuvre,* but he was neither a painter nor a filmmaker, and psychoanalysis was not one of these great arts, in which extravagance can be endlessly taken up and mastered by the overwhelming force of beauty.

Notes

1. Nicolas Abraham, Maria Torok, *Cryptonomie, Le Verbier de l'Homme aux loups,* with a preface, "Fors," by Jacques Derrida, Aubier Flammarion, 1976.
2. *Ornicar?,* no. 14, pp. 4–9.
3. An allusion to the fact that Nicholas Abraham and Maria Torok were members of this Society.
4. We shall overlook the fact that this is not Freud's position; he distinguished clearly, for example, between the dream processes he had invented and the actual genesis of the dream, and emphasized the sexual motivation of the joke. But Jacques Lacan's return to Freud does not go into these details.
5. "I pointed out that the knowledge in question was neither more nor less than the unconscious. It was very difficult to know for sure just what Freud had in mind. But what he says about it required, it seemed to me, that it be a kind of knowledge." ibid., p. 5.
6. Lacan's enterprise inevitably reminds you of the man who is looking for his key at the foot of a gas lamp. To a passer-by who asks him whether that was where he lost it, he replies: "No, but there's more light here."
7. Discreetly, but insistently, Octave Mannoni had long since drawn attention to this fundamental point. Cf. notably his collection of articles, *Clefs pour l'imaginaire ou l'Autre Scène,* éditions du Seuil, 1969.
8. On this question, which philosophers will be familiar with, but which psychoanalysts cannot turn their backs on, see Michel Henry's book, *Généalogie de la psychanalyse,* P.U.F., 1985.
9. *Ornicar?,* no. 25, p. 23.
10. *Revue philosophique de Louvain,* November 1979, p. 564.

11. Ferdinand Cheval (1836–1924), postman in the village of Hauterives to the west of Grenoble, spent some 33 years constructing his "Palais Idéal." Part temple, part monument, part palace, and never intended as a dwelling, Cheval's edifice is a weirdly repulsive yet also intriguing work of art, a fantastic agglomeration of styles, materials, and cultures. [trans.]

Glossary

The contents of this book do not assume any previous acquaintance with Lacan's *oeuvre;* the reader will find sufficient explanation of the latter in the text. However, for those who are not totally familiar with the *oeuvre,* it seemed useful to provide a glossary, giving the origin and meaning of Lacan's major notions. The symbol * denotes a separate entry for the term in question.

Borromean knot In the final phase of his work, Lacan frequently appeals to notions drawn from topology (the Möbius strip, Klein's bottle, etc.) and particularly from the mathematical theory of knots, with a view to systematizing the relations between the Real, the Symbolic, and the Imaginary. The Borromean knot is a figure found in the coat of arms of the Borrome family, consisting of three interlocking circles, joined in pairs, as in the Olympic symbol.

Castration Freud had discovered the importance of the threat of castration in his analysis of phobias, in particular, the case of Little Hans. He subsequently made it an essential feature of the Oedipus complex, as well as the cause of anxiety.

For Lacan, castration is a metaphor* for what the subject* lacks and must accept. It is the third term of a progression in his accession to truth and desire. At first, the subject experiences *deprivation,* quite simply because he is materially missing something; he is deprived of food, freedom of movement, or of love. In this state of deprivation, he experiences *frustration.* This thing of which one is deprived has been withheld by someone else. Frustration plays an important role in analysis. Indeed, if one wants the analysand to have access to his truth (see subject*), or to his desire, it will necessitate not responding to his demand.* The analysand will then be frustrated, for example, by the silence of the analyst.

But one cannot remain at this stage; access to castration implies that the subject recognizes that he will never attain the object of his desire. He is castrated because the phallus,* which represents all possible objects of desire and all signifiers, will never be his.

Castration is said to be imaginary when the subject thinks he can make up for his inadequacy by appropriating phallic power. It is said to be symbolic when the subject abandons his illusions of omnipotence. Symbolic castration is thus the aim of analysis.

Demand Demand is conceptualized in Lacan's work only in so far as it is related to need and desire.* The mother, or any person who takes the place of the mother, can, to begin with, immediately satisfy the needs of the newborn. But as the child grows, in order to satisfy its needs, it must, in one way or another, formulate a demand. In order to do so, it has to learn its native tongue. If needs have to be satisfied by being routed through language, language thereby forces them to be deferred. Repression stems from need having to pass through the signifier* in order to make itself understood by the other, and this necessity (being understood by the other) becomes a preoccupation which detaches itself from what is actually demanded. Demand centers on the person capable of satisfying needs, independently of these needs: This is the demand for love. The refusal to respond to this demand will precipitate the emergence of desire.* For a Lacanian, what is here described genetically merely serves to unveil a permanent structure characteristic of the speaking being, which is, for him, the definition of the human being.

Desire This notion has two origins: It takes up Freud's *Wunsch,* that is, the desire that dreams tend to satisfy, as well as Hegel's *Begierde,* that is, the life-force. First, Lacan overlooks the link with the state of sleep and the multiple forms it is able to assume; all he retains is that desire is characteristic of the unconscious, and that it is unrelated to objects in the external world. Second, Lacan forgets that desire is an elementary manifestation of life, so as to turn it into the loftiest expression of the truth of the subject.*

Desire is defined by its cause, the object *a,* that is, by lack and the void. It appears when the mother or the alter ego refuses to respond to the demand for love, whereby demand sees the object it was seeking disappear, and finds itself confronted with the void. It is this void that transforms demand into desire, and causes desire. In the analytic cure, the analyst's refusal to respond to the analysand's demands is supposed to provoke the emergence of desire, and the patient's accession to his desire; this is the aim of analysis.

It is difficult to reconcile this conception of analysis as accession to desire proper with the definition "The desire of Man is the desire of the Other," where desire finds itself alienated either in the desire of the other (in the imaginary relation), or in the desire of the Other (in the Symbolic), since the Other, which is an agency, is then itself supposed to desire. It is in this same perspective that the Other becomes the Law* and gives rise to desire.

Foreclusion See "Psychosis."

Imaginary See "Symbolic."

Frustration See "Castration."

Jouissance The word was borrowed from Georges Bataille, who used it to designate the link between eroticism and mysticism. The state of the fullest self-abandon is seen as inseparable from extreme sexual pleasure. This origin of the word appears in Lacan, for example, when he comments on the ecstasy of Bernini's Theresa of Avila (see the cover of vol. XX of the seminar, *Encore*). But for Lacan the word also covers what Freud referred to as being beyond the pleasure principle, where Eros and Thanatos meet up.

With *jouissance,* we are in fact beyond the principle of pleasure-unpleasure, distinct affections that point to what one can actively seek or flee. *Jouissance* is an indistinguishable mixture of extreme suffering and extreme pleasure. For example, when you place yourself in the position of victim, it is reasonable to wonder whether suffering is not accepted or even actively sought after, because of the *jouissance* it produces. But *jouissance* is not confined to the seriously pathological cases encountered in psychoanalysis; it is found in every patient as the final justification for clinging to his symptoms. The patient seems attached to his suffering, as if it were the price to be paid for his existence. *Jouissance* appears as a diffuse, permanent form of the orgasm, an orgasm achieved outside the sexual relation. Lacanian psychoanalysts turn to the concept of *jouissance* in order to conceptualize phenomena such as feminine sexuality, repetition, masochism, and the interminable aspect of analysis.

The Law Ordinarily, this word is used in the plural, and is distinguished from norms, rules, commands, and orders. In Lacanian discourse, these differences are blurred, and "Law" is used in the singular. It then designates the law of the signifier,* to which the subject is subordinated.

The Law is then related to desire. Taking up the dialectic of Saint Paul, who turned the Law into what exposes sin, Lacan posits the Law as giving rise to desire,* and desire as running up against the Law.

Matheme Invented by Lacan, this word is a transposition of Levi-Strauss's "mytheme." Levi-Strauss wanted to reconstruct the totality of myths by combining the smallest distinct elements, uncovered in the analysis of certain myths. Lacan's dream was to create an algebra that would recompose, and account for, all the complex realities encountered in analytic theory and practice, on the basis of a limited number of indivisible elements. Lacan was able to use letters to represent certain relations discovered by other means. But the dream remained just that—a dream, and so far no one has been able to provide a single example of a matheme.

The Name of the Father Psychoanalysis discusses the function of the Father in terms of the Oedipus complex. Moreover, Freud attributes a decisive importance to the primordial identification with the Father, in the constitution of the individual, or to the murder of the Father as the precondition for social ties. For Lacan, it is not the real father who matters, but the Symbolic father who intervenes through the name, that is, through the way in which the paternal function is stated in society. The subject's fate will depend on the place of the Name of the Father in the signifying chain constituting the subject's unconscious.

Symbolic castration* is supposed to operate through the introduction of the Name of the Father. In the case of psychosis,* the Name of the Father has not been taken into account, and has undergone foreclusion.

Object A The letter "a" appears in the diagram in which Lacan sums up his conception of the subject:* It represents the *other.* The term *object a* takes up the notion of the part-object established by Freud. Part-objects are objects separated from the body and invested with erotic interest.

To the list of the three part-objects (the breast, feces, and the phallus), Lacan adds the voice and the gaze. What he has in mind is no longer the libidinal development of the individual, but the structure of the subject.* As far as these objects are concerned, what Lacan retains is that they are prone to being lost, that they are, or can be, missing. The *object a* then becomes the fundamentally lost or missing object, the representative of lack. Lacan uses it to interpret or reinterpret those notions that seek to explain a problematic relation to objects. For example, the *object a* will become the cause of desire,* and the drive is reduced to a circular movement around it. The fantasy is then the multiform relation the subject maintains with this object, and primary narcissism will be the subject's first internal deviation, from which this object will emerge.

The Other There is nothing in Freud's texts that is related to this notion. On the other hand, the problem of the alter ego was discussed by the German and French phenomenologists: Hegel, Scheler, Sartre, Levinas, and so on.

What is original in Lacan's work, who places himself in this tradition, is that he does not distinguish between the ego and the alter ego. For him, the ego is defined precisely by the fact that it is caught, from the outset, in the image of the alter ego. Every ego is merely the reflection of the alter ego, which Lacan calls the *petit autre* [the other with a small "o"]. The ego resembles [est semblable à] the alter ego, and models itself on it, especially in relations of rivalry, jealousy, and prestige.

By contrast, the Other [le grand Autre] is the other revealed in speech, prohibitions, and orders. It is not a counterpart [un semblable], but an entity that lays down laws, governs exchanges, and presides over the circulation of the Symbolic.*

The Other ends up meaning (it is thus not any person in particular) the bearer of language, the locus of the fundamental signifiers. But certain figures are able to occupy the position of the Other: The mother, insofar as she represents paternal discourse; the father, insofar as he represents the Law* and represses desire;* or the teacher, who transmits cultural and social values.

In Lacanian writings, the Other will occasionally be "barred" or "crossed out," meaning that it is distinct from any existing person.

The Pass The procedure of the pass (which refers to the passage from the status of analysand to the status of analyst) went as follows: A beginner-analyst who wanted to take the pass would draw the names of two analysands, called "passers" (who were themselves undergoing the pass—that is, who were in the process of becoming analysts, and who had been chosen by their analyst). He would go and talk to them, for as long or as briefly as he liked, about his analysis, and especially about everything to do with the fact of his becoming an analyst. The passers would then present themselves before an "acceptance jury" (initially appointed by Lacan, and subsequently elected by the members of the École Freudienne de Paris) and would present the results of the pass and their "passee," whereupon the jury would decide whether the "passee" would receive the title of analyst of the School, the highest rank in the hierarchy.

Phallus Freud used this word to designate the third phase in the development of the human individual. The phallus is the symbol of the force of virility. Whereas the penis is the natural organ, a part of the body, the phallus is an independent symbol, which is able to circulate, or which can be hidden or concealed.

The term allows Lacan to link his own theory of the signifier to the Freudian theory of the libido; in it, language and sexuality are joined together. For reasons that are never made explicit, reasons dictated by the requirements of Lacanian systematization, the phallus becomes the signifier *par excellence,* and all other signifiers in the language are dependent on it. As detached from the body, it also becomes the symbol of castration,* and is marked with a minus sign, meaning that it also symbolizes lack.

The imaginary phallus becomes the object of imaginary castration, which maintains the illusion of being able to satisfy desire. The symbolic phallus is referred to symbolic castration, which makes accession to desire* possible.

Privation See "Castration"

Psychosis The nosographical classification varies from country to country. In France, in part following Freud, psychoses, neuroses, and perversions are currently differentiated. Taking up Freudian terminology in his own way, Lacan distinguishes between neurosis (marked by repression—*Verdrängung*—and especially characterized by the refusal of symbolic castration*), perversion, defined by the denial (*Verleugnung*) of castration, and psychosis, defined by the foreclusion (*Verwerfung*) of the Name of the Father.*

Foreclusion shows us why the psychotic finds symbolization impossible: He lacks the necessary means of using words in such a way for them not to be taken literally.

The Real Lacan borrows this word from the epistemologists, who use it to designate what science discovers in the way of immutable laws underlying phenomenal reality. This notion, therefore, initially serves to designate the structures that can be separated out from imaginary relations. Later, the word designates what the psychotic, in his hallucinations, takes for reality. Since for him symbolization is impossible, he transfers symbols into reality; he sees them as real, as the Real.

The Real subsequently becomes what resists symbolization, the impossible to symbolize, which can be known only through the effect of a shock and about which nothing can be said, which is why the characteristics of the Real are all negative. The Real is a limit you run up against; it is the void, the impossible, an impasse.

Signifier Saussure had defined the linguistic sign as the indissoluble unity of the signifier (verbal materiality) and the signified (meaning). Lacan separates the signifer from the signified. Overarching the analysis of the dream, which he sees as a rebus, or the analysis of slips and jokes, which play on the plurality of meaning, the same signifying material is supposed to be able to support various meanings. It is difficult to furnish examples without immediately resorting to puns or spoonerisms; nonetheless, the most famous example is provided by the title of one of the seminars: Should it be understood as "les Noms du père" [the Names of the Father] or as "les non-dupes errent" [non-dupes err]?

Lacan based his theory on Jakobson's study of aphasia, of which two kinds were discovered—the first bearing on substitution, where paradigms are affected, and the second bearing on combination, where syntagms are affected. Jakobson concludes that, in order to speak, we make use of two functions, which he calls metaphor and metonymy.

Lacan follows up what he finds suggestive about these two functions, applying them to the two dream mechanisms elicited by Freud: Condensation and displacement. In the unconscious, we then find nothing but signifiers, linked together in signifying chains; symptoms are formed through metaphor, and desire is a metonymy. Thus, Lacan believes he has shown that the unconscious is governed by the laws of the signifier,* and that he has forged the connection between the science of dreams and linguistics, and is therefore able to claim that the unconscious is "structured like a language."

Subject In Lacan, the notion of subject is difficult to grasp, because it stems from an oscillation between a critique of the philosophical conception of the subject, a critique basing itself on Freud, and a critique of Freud (who has no concept of subject) that is based on linguistics.

The classical philosophical problematic thinks the subject in relation to the object; the two terms are correlative. The French psychoanalyst M. Bouvet took up this conception when he elaborated the notion of object relation in order to present the doctrine of the three stages (oral, anal, phallic).

Lacan rejects this conception, since for him these part-objects are to be defined as lost objects, and he groups them under the term *object a.** It is impossible to constitute a subject, in the traditional sense, in opposition to this object, which is always already missing.

The true subject will be the subject of the unconscious. Commenting on the Freudian formulation: "Wo es war, soll ich werden," and contrary to the current interpretation, which has it that the Id gives way to the Ego, Lacan maintains that, thanks to analysis, the Ego loses its (illusory) place in favor of the Id, which is the truth of the subject.

Lacan's answer to the age-old question "Who am I?" would be: I have to become the Id. But this is subject to the condition of not conceiving this latter, as Freud does, in dynamic and energetic terms. The "Es," which in Lacan becomes the "S," is merely the effect of the signifier,* since, in the unconscious, there are nothing but signifiers. The subject has no consistency in itself; it is merely the "passion of the signifier."

But we can ask ourselves whether this isn't simply a repetition of the traditional conception of the subject in another form: We have a subject corresponding to a missing object, a subject that, since it vanishes in the play of signifiers, in fact resembles its object.

Symbolic In Lacan's work, this substantive has a double origin: First, it stems from the adjective "symbolic," used mainly by ethnologists in order to designate the systems that make the exchanges at the core of a society possible. Second, it refers to the notion of the symbol used in algebra. In identifying the two senses of the term, Lacan had provided himself with a universal principle of functioning, which was supposed to allow him to comprehend man, the speaking-being, and subject him to the calculable.

The Symbolic is inseparable from the Imaginary and the Real.* The Imaginary is defined by the relation to the counterpart [semblable] (see *Other*)—that is, by the relation to the dual confrontation typical of neurosis. In current Lacanian discourse, the Imaginary has pejorative connotations, as in the expressions: "to remain stuck in the Imaginary," "all that is merely imaginary." By contrast, the Symbolic is endowed with a positive connotation, since psychoanalysis is supposed to be a movement from the Imaginary to acceptance of symbolic castration.*

The Symbolic relates to the Real as its limit: The Real is what the Symbolic runs up against, what resists it.

Name Index

Subject Index